Logistics

Series Editor
Jean-Paul Bourrières

Logistics

Joëlle Morana

iSTE

WILEY

First published 2018 in Great Britain and the United States by ISTE Ltd and John Wiley & Sons, Inc.

ISTE Ltd
27-37 St George's Road
London SW19 4EU
UK

www.iste.co.uk

John Wiley & Sons, Inc.
111 River Street
Hoboken, NJ 07030
USA

www.wiley.com

Library of Congress Control Number: 2018941063

British Library Cataloguing-in-Publication Data
A CIP record for this book is available from the British Library
ISBN 978-1-78630-310-3

MIX
Paper from responsible sources
FSC www.fsc.org FSC® C013604

Contents

Introduction

This book is comprised of 42 complementary sheets exploring logistics. Each sheet provides a key point, in-depth analysis and dedicated references.

I.1. Sheet 1: The Logistics Function

Key point

The logistics function is central in and between companies. Some even say that without logistics, there can be no professional activity. The importance of its evolution is reflected throughout the 20th and 21st Centuries. Generally speaking, there are five key dates. Before the war of 1939–1945, we were in a "stewardship" rationale. We managed production activity as best we could. From 1945 to the 1970s – the period of the Glorious Thirty – attention was paid to the physical flow and in particular the downstream flow of distribution. We were in a "mass consumption" phase, where the company's interest was to deliver as many products as possible to the maximum number of customers. We were in a "standardization" rationale here, in an approach known as "push flow". Industrialization and competition in the 1980s forced companies to tightly control their costs. This financial control was part of logistics activity, through the implementation of analytical accounting (often evaluated using Activity-Based Costing) which allowed us to assess the profitability of a product and/or a range of products. The 1990s highlighted the "flow of information" as a key issue in logistics. By this, we mean the importance of "customization", a "pull-flow" approach where the customer's "fixed and definitive order"[1] is the one that "triggers" production. This period would underscore the shift from a "logistics strategy" (support for strategy) to a "strategic logistics" (strategy foundation) [FAB 94]. At the dawn of the 21st Century, we refer to "sustainable supply chain management", in the sense of sustainable development. The economic, environmental and social/societal roles of logistics are highlighted here [MOR 13].

I.1.1. *Definitions related to the supply chain*

Several definitions are linked to the supply chain. These include ASLOG (*Association française de la Supply Chain et de la Logistique*) and CSCMP (Council of Supply Chain Management Professionals).

For ASLOG, the supply chain is defined as "*the global supply chain, which goes from supplier to customer and where production is demand-driven. Its objective: the right product in the right place, at the right time*". For CSCMP, the supply chain is apprehended "*1) starting with unprocessed raw materials and ending with the final customer using the finished goods, the supply chain links many companies together; 2) the material and informational interchanges in the logistical process stretching from acquisition of raw materials to delivery of finished products to the end user. All vendors, service providers and customers are links in the supply chain*".

Nevertheless, only the definition of the AFNOR standard NF X 50-600 (1999) prevails. It is not about supply chain, but logistics. In this regard, it can be said that "*logistics is defined as the planning, execution and control of the movements and implementations of people, goods and support activities related to these movements and implementations, within an*

1 All purchases made within the European Union are fixed and final, which means that you cannot withdraw unless you pay a cancellation fee. Nevertheless, law no. 2014-1545 of March 17, 2014, known as the *Hamon law*, in Article L. 121-21, paragraph 1, of the French Commercial Code specifies that "*The consumer has a period of fourteen days to exercise his right to withdraw from a contract concluded at a distance, following a telephone or off-premises solicitation, without having to justify his decision or bear any costs other than those provided for in Articles L. 121-21-3 to L. 121-21-5. Any clause by which the consumer gives up his right of withdrawal shall be null and void*". But beware; some companies circumvent this article by having the consumer validate a purchase of goods and services in a computational manner that cancels this law when this is specified in the General Terms and Conditions of Sale.

organized system to achieve specific objectives. [...] The purpose of the logistics function is to satisfy expressed or latent needs at the best economic conditions and for a given level of service".

Seven main stages are assigned to the logistics process by the experts of this standard (derived from [GIA 03]):

– "identify *the market needs in terms of quality of service and determine the objectives for quality of service*;

– design *the logistics system and the organization of the flow chain; define the logistic characteristics of the product (needs and constraints), those of the after-sales system*;

– develop *the logistics system, organization, procedures and logistic information systems, after-sales service system, packaging and guarantee the availability of operational resources*;

– production, *in other words, implement industrial processes and systems, plan and schedule material and service requirements, procurement*;

– selling, *in other words, implement distribution procedures and systems, storing, packaging, dispatching, transporting and installing products; control the execution of transport and distribution operations; managing returns*;

– support, *in other words, implement maintenance procedures and systems, repair and distribution of parts; acquire, store, package, ship, transport and deliver parts; maintain and repair products; recover and recycle products*;

– control *the performance of the logistics system: exploit the results, compare them with objectives, make corrections, anticipate...*"

I.1.2. *Definitions related to Supply Chain Management*

Since the 1990s, the term "Supply Chain Management" has been gaining popularity. Supply Chain Management is the management of global logistics. This global management underlines the strategic, intra- and inter-organizational role of logistics. One of the main differences between supply chain and Supply Chain Management is that in Supply Chain

Management, there is a clear desire to establish long-term partnerships in order to share gains and losses.

It was through the writings of Martin Christopher [CHR 92] that the notion of Supply Chain Management (SCM) took off. At the time, he defined SCM as:

> *"The supply chain is the network of organizations that are involved, through upstream and downstream linkages, in the different processes and activities that produce value in the form of products and services in the hands of the ultimate consumer"* [CHR 92].

Several other definitions followed, the most referenced of which are the following:

> *"There is a need for some level of coordination of activities and processes within and between organizations in the supply chain that extends beyond logistics"* [LAM 98].

and

> *"The systemic, strategic coordination of the traditional business functions and the tactics across these business functions within a particular company and across businesses within the supply chain, for the purposes of improving the long-term performance of the individual companies and the supply chain as a whole"* [MEN 01].

I.1.3. *Definitions related to Sustainable Supply Chain Management*

Currently, the focus is on Sustainable Supply Chain Management. All of the research reveals a wide range of application areas for implementing sustainable logistics. Works can be found on waste management, green logistics, reverse logistics, eco-design and the reduction of greenhouse gas (GHG) emissions through rationalization and massing of transport routes, particularly on the logistics of the last mile [COL 03, BER 08, BAZ 08, BLA 08, DUR 10, GON 10, MAS 10, PAN 10, SAV 10, BLA 10]. However, sustainable logistics considerations are gradually converging towards the inclusion of the three economic/environmental/social–societal pillars, throughout the logistics chain [MOR 13].

In terms of the definitions of Sustainable Supply Chain Management (SSCM), we can quote:

> *"SSCM requires a broadened approach to SCM. It should emphasize economic, ecological and social aspects of business practices and theory."* [SVE 07];

> *"Sustainable Supply Chain Management (SSCM) is the strategic, transparent integration and achievement of an organization's social, environmental and economic goals in the systemic coordination of key inter-organizational business processes for improving the long-term economic performance of the individual company and its supply chains."* [CAR 08];

> *"We define Sustainable Supply Chain Management as the management of material, information and capital flows as well as cooperation among companies along the supply chain while taking goals from all three dimensions of sustainable development, that is, economic, environmental and social, into account which are derived from customer and stakeholder requirements."* [SEU 08];

> *"A sustainable supply chain is then one that performs well on both traditional measures of profit and loss as well as on an expanded conceptualization of performance that includes social and natural dimensions"*; *"If a sustainable chain is one that performs well on all elements of the triple bottom line, Sustainable Supply Chain Management is then the specific managerial actions that are taken to make the supply chain more sustainable with an end goal of creating a truly sustainable chain."* [PAG 09];

> *"Sustainable Supply Chain Management is valued as the management flows of material, information, financial, people and intelligence for economic, environmental and social/societal purposes. From a strategic management approach, it is driven by the voluntary interplay of intra- and inter-organizational connections and the long-term performance of each company and its supply chain."* [MOR 13].

I.2. Sheet 2: Supply Chain Flows

Key point

The supply chain is differentiated from Supply Chain Management by the degree of cooperation between the different partners in the channel. In a very schematic way, in a supply chain we look for effectiveness, that is, we (simply!) answer the customer's order. In Supply Chain Management, efficiency is sought between partners so that the order is carried out using as few resources as possible, and also in a spirit of sharing gains and risks. But whether we talk about supply chain or Supply Chain Management, there is always a flow history behind it.

I.2.1. *Fundamental flows in the supply chain*

As it evolved, logistics saw an increase in the number of flows associated with its practice. Basically, logistics refers to two main flows [HES 73, TIX 79, COL 88, FAB 97, ARN 98, PAC 99]:

– a product flow (a term sometimes replaced by material flow, goods flow, physical flow and/or service flow);

– and a flow of information.

These first two flows maintain links upstream and downstream of the supply chain.

Aurifeille *et al.* [AUR 97] integrated the cash flow into these two basic flows, with the aim of short- and medium-term profitability. This flow goes from downstream to upstream, and also in the opposite direction.

The supply chain is thus generally represented by these three flows:

– originally, a flow of products from upstream (the supplier) to downstream (the final customer). Notwithstanding, the flow of products can also have an inverse direction, that is, from downstream to upstream for varying (1) commercial reasons (technical problems and trade repossessions), (2) legal reasons (recycling) and (3) economic reasons (re-use and recovery) ([LAM 04], see Sheet 35, section 6.3.4).

– the flow of information from downstream to upstream, because it reflects the description of the command;

– the cash flow also from downstream to upstream, which corresponds to the amount of the purchased good, except in the case of reimbursement.

However, it should not be forgotten that the fulfillment of an order requires a fourth type of flow, namely the people flow [HES 73, TIX 83], without which the existence of the first three flows could not exist!

Finally, a final type of flow is suggested, in order to express the functioning of a supply chain, namely the intelligence flow [MES 99]. Through the basic logistics triptych "cost – quality – time"[2], these two authors argued over an intelligence flow, with the aim of maximizing the exploitation of all information, in order to optimize 1) *responsiveness*, in the sense of identifying and then satisfying unforeseen demands; (2) *agility*, in other words, the ability to rapidly reconfigure an offer system by redeploying available resources and (3) *efficiency*, in the ability to systematically eliminate all forms of waste.

I.2.2. *"Internal" flow policies in the supply chain*

In logistics, reference is also made to three types of internal flows that can be found in production management. These flows are applied according to the company's policy and also according to the competitive level of the activity sector in which the company is located. For example, a company that is part of a "niche" strategy, where expensive products are generally found with a selective clientele, can more easily apply a pull-flow or tight-flow policy.

The three "internal" flow policies in production that can be used are:

– push system. This type of production is practiced for standard goods in bulk. The client's order is not expected. Based on sales forecasts, the company determines a production stock and sends it to distributors for a

2 The representation of logistics is established by a triangle in which the three elements to which that logistics must seek to respond are specified: good cost management, product and service quality and a respected deadline.

supposed sale. Thus, in this particular context, information flows go in the same direction as product flows. The constraints of this type of production turn out to be the presence of large stock with a risk of losses, in the event of obsolescence and/or non-sale of products. In addition, the presence of stock also means costs in terms of holding stock and warehouse management;

– pull system. The production process is triggered only when the order is fixed and definitive, so at the customer's clearly identified request. The production order starts from the purchase order. We are looking here at a personalization of the production. However, intermediate stock may exist in a pull flow production;

– tight-flow policy. This is often assimilated to a pull-flow policy, since it is also based on fixed and definitive control. It is part of the "5 zeros" method, which presupposes a management without defects, paper, breakdowns, stock or delay. It is similar to the just-in-time[3] or lean manufacturing method[4]. In current practice, a tight-flow policy aims at production without stock throughout the cycle.

For companies confronted with competition, it is almost impossible for them to have a policy based solely on pull and/or tight flows. Companies must therefore respond to the dilemma of a search for economic profitability, by minimizing inventory costs while proposing a customization of their order, in order to increase customer loyalty, and all this while within the delivery time. This often takes the form of a combined push-flow–pull-flow policy. Upstream, we standardize products as much as possible with the setting up of stock, and when the order arrives, we customize the product. This approach is apprehended under the concept of "*delayed differentiation at the production level [... which] delays changes in the form and identity of products and production processes*" [GAR 08, p. 32].

---

3 For Bigras *et al.* [BIG 95], just-in-time is approached through four fundamental principles: (1) producing in exact quantities; (2) producing unitarily; (3) respecting individuals and (4) holding a long-term outlook. The objective is threefold: elimination of waste, constant improvement and elimination of unforeseen events (see also Sheet 25, section 4.3.3).

4 "*Lean manufacturing is a generic name for a production system originally developed by Toyota, and now used worldwide in all industrial sectors. Lean is defined as a systematic approach that aims to identify and eliminate all waste (non-value activities added) through continuous improvement to achieve industrial excellence*" [PET 06, p. 3].

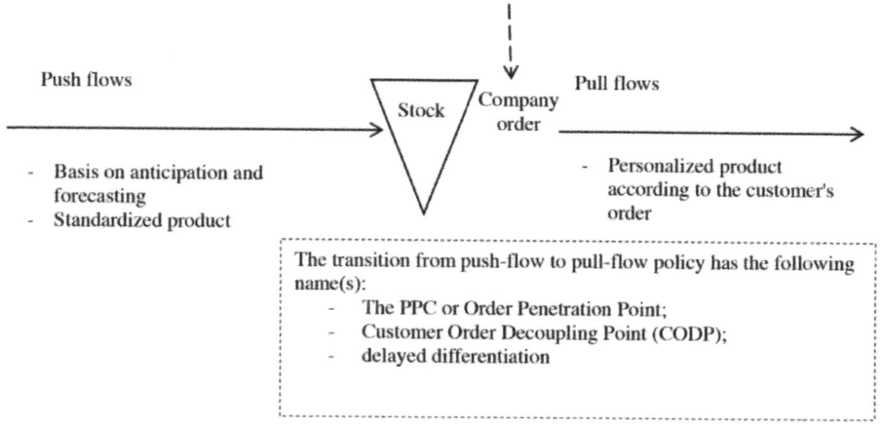

Figure I.1. *The combination of push- and pull-flow production*

I.3. Sheet 3: The Main Models of the Global Supply Chain

Key point

Martin Christopher [CHR 82] was the first author to point out the importance of Supply Chain Management (SCM) and its modeling. Several studies have subsequently developed new models. As a reminder, some definitions of SCM are proposed in Sheet 1.

I.3.1. *The Supply Chain Management structure according to Christopher (1992 onward)*

Martin Christopher [CHR 92, CHR 94, CHR 97, CHR 99, CHR 00] was one of the first authors to highlight the importance of Supply Chain Management in logistics practices. From these early writings, we will retain the aspects presented in Figure I.2.

For Christopher [CHR 00], the notion of partnership is important for the practice of SCM. In order for a partnership to be set up, there must be a combination of agility, reliability, responsiveness and proactivity. For this author, agility implies "*the ability to move quickly and to meet customer demand sooner*". Reliability in logistics depends on the ability to deliver to the right place, at the right time, on the right date and to the right customer. The difference between responsiveness and proactivity is that the former identifies and responds quickly to changes in its environment, while the latter responds quickly when proactivity influences changes in that environment.

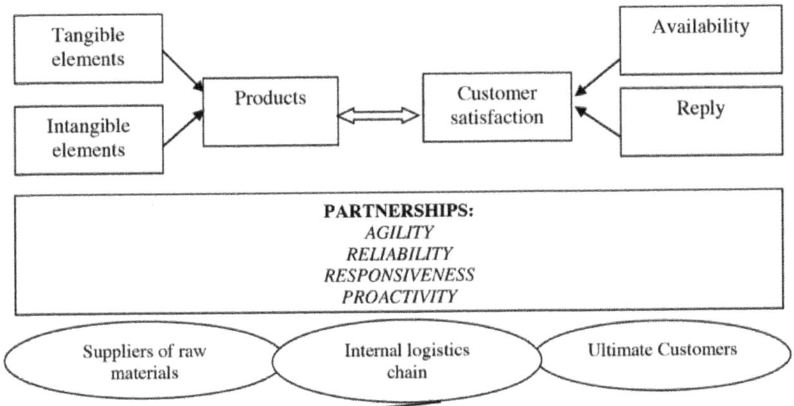

1. PROCESS	- innovative products/brands - supplier management - management of the supply chain - prospective development
2. PERFORMANCE	- customer satisfaction - flexibility - personal commitment
3. CREATING VALUE FOR CUSTOMERS	
4. VIRTUAL INTEGRATION BETWEEN SUPPLIERS	- information sharing - creation of common strategic objectives

Figure I.2. ***The SCM structure*** *(source: adapted from [CHR 92, CHR 94, CHR 97, CHR 99, CHR 00])*

In addition, for SCM to succeed, internal actions are needed. As such, four elements must be interlinked: good process management, performance measurement, value creation for ultimate customers (that is, downstream logistics) and integration with suppliers (that is, upstream logistics).

I.3.2. *The World Class Logistics model [GLR 95, BOW 99]*

The WCL model (World Class Logistics) [GLR 95, BOW 99] advocates four areas of competence. These four areas of skill are themselves divided into several success criteria. These areas of competence are:

– strategic positioning with four criteria, namely *logistics strategy* (financial and commercial objectives), the *supply chain* (synchronization of resources), *infrastructure/networks* and *organization of people*;

– integration, which considers seven criteria: (1) *unification of the supply chain*, (2) *information systems*, (3) *information sharing*, (4) *compatibility to be exchanged*, (5) *standardization of policies and procedures*, (6) *simplification of procedures* and (7) *people's support*;

– agility through the criteria of *vigilance*, *adaptability* and *flexibility*;

– performance measurement with three criteria: the *choice of indicators*, *evaluation of the supply chain process* and *benchmarking* techniques.

I.3.3. *The SCOR® Supply Chain Operations Reference Model (1996)*

Introduced in 1996, the SCOR model is the product of the Supply Chain Council, a group of consulting companies of Pittiglio, Rabin, Todd and McGarrah, and AMR[5], as well as 69 initial companies from various economic sectors (more than 400 companies currently registered). The SCOR model is structured into four levels:

– level 1 (Top Level) defines the perimeter of the model. The five key management processes, namely planning, procurement, fulfillment, delivery and return management are presented here (see Sheet 41, section 7.6.3 for a presentation of generic metrics);

– level 2 (Configuration Level) specifies the strategic processes that companies want to implement within their supply chain;

– level 3 (Process Element Level) breaks down each initiated process (input and output elements, metrics, best practices);

– level 4 (Implementation Level) is at the initiative of each company. It specifies level 3 according to the desired interests in a more or less long-term perspective.

5 The Supply Chain Council was formed in 1996–97 on the initiative of PRTM, AMR Research, Procter and Gamble, Texas Instruments and other organizations. PRTM: Pittiglio, Rabin, Todd and McGarrah is a consulting company created in 1976 with the aim of providing technological assistance to various international companies. AMR: Advanced Manufacturing Research, created in 1986, is a consulting company specialized in the programming of process management measurement systems.

I.3.4. *The model of the Global Supply Chain Forum Structure of [LAM 98]*

The structure of Supply Chain Management according to Lambert *et al.* [LAM 98] is broken down into several elements:

– the first element identifies the members of the supply chain, considering those who are in the *primary order* of the productive process and *secondary order* in the supply of goods and services;

– the second element evaluates the aspects of the network of organizations, in other words, the number of operators on the *horizontal* chain, the number of operators on the *vertical* level at each horizontal "node" and the *positioning of the company* on the processed logistics chain;

– the third element considers the level of integration of the company in its supply chain (power of influence);

– a fourth element identifies and analyzes eight business processes in the supply chain: (1) customer relationship management, (2) customer service management, (3) demand management, (4) fulfillment of orders, (5) workflow management, (6) supplier relationship management, (7) product development and marketing, and (8) return management. These eight processes are "linked" to each other in terms of the *functions* present in the company (marketing, sales, research and development, logistics, production, purchasing and finance) as well as to other operators in the supply chain (suppliers and customers);

– a fifth element lists the *physical* and *technical* management components (planning and control methods, work rhythms and activity structure, organization structure, communication and information flow of the technical structure, product flow of the technical structure) as well as *managerial* and *behavioral* components (management methods, power and leadership structure, risk and reward structure, culture and attitude).

I.3.5. *The model by Mentzer et al. [MEN 01]*

Mentzer *et al.* [MEN 01] suggested two projects to structure Supply Chain Management:

– Supply Chain Orientation (SCO), which encompasses six principles, namely (1) *credibility* (reliability, recognition of the company),

(2) *benevolence* ("health" of the company, answers to questions, risk sharing and rewards), (3) *commitment* (cooperation, technical assistance), (4) *support for Top Management* (dissemination of objectives, long-term contracts, training), (5) *compatibility* (strategy, general knowledge, etc.) and (6) *standards* (cooperation, value creation).

– Supply Chain Management (SCM), which also involves six principles: (1) *common vision and goals* (standardization of practices, definition of roles), (2) *information sharing* (Electronic Data Exchange, forecasting and planning exchange), (3) *risk sharing and rewards* (financial, research and development), (4) *cooperation and integration of processes* (reporting, indicators, quality, communication, stock management, etc.), (5) *long-term relationships* (reducing complexity, audits) and (6) *supply chain leadership* (audit, benchmarking).

In 2004, Min and Mentzer completed these two projects with a third, which developed a "performance" axis with five principles to consider: *availability* (inventory), the *offer of products and services*, *timeliness* (just-in-time delivery), *profitability* and *growth* (turnover and market share).

I.4. Sheet 4: The Main Logistics Associations

Key point

Several associations on transport and/or logistics exist at local, national and international levels. They provide an outlook on the subjects dealt with by professionals and/or researchers in this field. Their objective is to ensure maximum dissemination of information. The list proposed presents the main organizations.

I.4.1. *The main national and international associations of logistics professionals*

I.4.1.1. *AFILOG*

AFILOG is a French professional association created in 2001. Its objective is to "*answer the questions of its members on the problems of a sector that is both rapidly developing and changing, composed of a mosaic of professions whose universe is often complex to understand*". Seven working themes are to its credit: the sustainable development commission, which deals with the Norme Française Bâtiments Tertiaires – HEQ® process (High Environmental Quality), the AFILOG charter, the evaluation of the energy performance of existing buildings, green rating™ and technical innovations; the urban logistics commission; the communication commission; the taxation commission; the risk prevention and regulation commission; the training commission and the prospective commission.

http://www.afilog.org/

I.4.1.2. *L'Association française de la Supply chain et de la Logistique (ASLOG) (The French Supply Chain and LOGistics Association)*

Founded in 1972, ASLOG has had several names including the "*Association des logisticiens d'entreprise*" (Association of Company Logistics Specialists), the "*Association pour la logistique dans l'entreprise*" (Association for Company Logistics) and the "*Association française pour la logistique*" (French Association for Logistics). Now known as the French Supply Chain and Logistics Association, ASLOG "*aims to promote supply chain and logistics professions on a national and international level. Multi-sectoral and representing more than 600 companies, it is the only association covering all activities in the global supply chain*". Currently, 11 thematic commissions are on the agenda within this association: benchmarking, the reading committee, skills/jobs and salaries, the e-supply chain, real estate and trends, health logistics, social and solidarity logistics, urban logistics, risk management, regulations in real estate and agile supply chain. ASLOG is also affiliated to the European Logistics Association (ELA).

https://www.aslog.fr/

I.4.1.3. *Le Bureau de Promotion du Shortsea Shipping (BP2S) (Shortsea Shipping Promotion Office)*

The BP2S is an association under the law of 1901, which was created in 2000. Its role is to promote shortsea shipping and intermodality. It is part of the

ESN (European Shortsea Network), which aims to promote the shortsea shipping market.

http://www.shortsea.fr/

I.4.1.4. *The Council of Supply Chain Management Professionals (CSCMP)*

Established in 1963 in the USA, the National Council of Physical Distribution Management (NCPDM) became known as the Council of Logistics Management (CLM) in 1985, and then as the Council of Supply Chain Management Professionals (CSCMP) in 2004. It is an international association with more than 8,500 members from industry, government and research sectors in 67 countries. It offers CSCMP's SCPro™ certification for the recognition of industrial supply chain expertise.

https://cscmp.org/

I.4.1.5. *The European Logistics Association (ELA)*

ELA is a European logistics association that provides an open link and forum for all logistics companies. Thusly, it proposes logistic standards to be used internationally. It is a federation of 30 associations covering Western and Central European countries (Austria: BVL Austria, Belgium: VIB and ABCAL, Croatia: HRLA, Cyprus: CLA, Czech Republic: CLA, Finland: Logy, France: ASLOG, Germany: BVL and BME, Greece: HILME, Hungary: HALPIM, Republic of Ireland: CILT, Italy: AILOG, Lithuania: LLA, Morocco: AMDL, the Netherlands: VLM, Norway: NIMA, Poland: PSML and PTL, Portugal: APLOG, Romania: ARILOG, Russia: NLA, Slovenia: SLA, Spain: CEL, Sweden: SILF, Switzerland: GS1 Switzerland, Ukraine: ULA, and the United Kingdom: CILT; http://www.elalog. eu/members).

http://www.elalog.eu/

I.4.1.6. *Global Standards 1 (GS1)*

Formerly Gencod, GS1 is a world organization created in 1973 under Belgian law. Its business is to create common standards for the identification of products, such as barcodes.

http://www.institutionnel.gs1.fr/Accueil

I.4.1.7. *Le Group of Terrestrial Freight forwarders (GTF)*

The GTF, created in 1987, is a French association created under the law of 1901. It brings together transport companies of Grouping-Messaging. It has developed messages of information exchange under the INOVERT® brand (International Overland Transport) and has also defined recommendations on the use of barcodes in the field of traceability. The GTF is a partner of GS1 (formerly Gencod), which is an organization founded under Belgian law, created in 1973.

http://www.institutionnel.gs1.fr/Partenaires-institutionnels/GTF

I.4.2. *The main clusters of French competitiveness in logistics*

I.4.2.1. *Nov@log*

Created in 2005, the Nov@log cluster of competitiveness aims to develop the services and logistics systems of the future. Members are divided into three associations: business, research and teaching, and government agencies. The cluster's topics of commitment are traceability, intermodality, the industrial logistics process and flows, the maritime and land interface, urban and logistical movements, the environment and security.

http://novalog.eu/

I.4.2.2. *LUTB Transport & Mobility Systems*

Created in 2005, LUTB Transport & Mobility Systems is a competitive cluster focused on the study of public passenger and goods transport in urban areas. Since June 2006, the LUTB-RAAC association has brought together the clusters of competitiveness and the Rhône-Alpes Automotive Cluster. LUTB-RAAC is an association under the law of 1901. LUTB has 185 members and is part of five research programs: (1) motorization and drive train, (2) safety and security, (3) vehicle architecture, (4) transport systems and (5) modeling and mobility management.

http://lutb.fr/

I.4.3. *The main French associations of teacher–researchers in logistics*

I.4.3.1. *L'Association Française des Instituts de Transport et de Logistique (AFITL) (The French Association of Transport and Logistics Institutes)*

AFITL is an association created in 1989, whose statutes were amended in 1998 in order to add logistics to its name. According to Article 1 of the statutes, the purpose of AFITL is to:

– "*publicize and ensure the promotion and development of research and higher education in transport,*

– *study common issues in these two areas,*

– *enable the representation of its members at the national and international level to the organizations concerned, by teaching and research activities in the fields of transport and logistics*".

Since 1994, AFITL has published the scientific journal *Cahiers Scientifiques du Transport*. Since 1995, it has also organized the Eric Tabourin doctoral days and awarded a prize for the best professional dissertations from Bac+4 to Bac+6.

http://afitl.ish-lyon.cnrs.fr/index.php/accueil.html

I.4.3.2. *L'Association Internationale de Recherche en Logistique (AIRL) (The International Logistics Research Association)*

AIRL was created in 2002. Its objective is "*to increase the visibility and dissemination of French-speaking research conducted in these fields* [logistics management and Supply Chain Management]". Every two years, it organizes the *Rencontres Internationales de Recherche en Logistique* (International Research Meetings in Logistics). It is supported by the scientific journals *Logistique et Management* and *Supply Chain Forum: an International Journal*.

http://www.airl-scm.com/

I.4.4. *France's leading transport and logistics research laboratories*

I.4.4.1. *Le Centre de Recherche sur le Transport et la Logistique (CRET-LOG) (Transport and Logistics Research Centre)*

Created in 1973 as CRET (*Centre de Recherche d'Economie des Transports*), this became CRET-LOG in 1992. It focuses mainly on three areas of research: (1) logistics and Supply Chain Management, (2) distribution channel management and (3) management of cross-organizational strategies.

https://cret-log.univ-amu.fr/

I.4.4.2. *The Laboratoire Aménagement et Economie des Transports (LAET)*

Created in 1979, the LET (*Laboratoire d'Economie des Transports*) is a research laboratory which specializes in transport economics and land development. The LET is attached to the CNRS, the University of Lyon 2 and the *Ecole Nationale des Travaux Publics de l'Etat* (*Ecole d'ingénieurs* ENTPE). The LET's work is in the fields of relations between transport (people and goods), territories and society. On December 10, 2015, the LET changed its name to LAET, short for *Laboratoire Aménagement Economie Transports*.

http://www.laet.science/

I.4.4.3. *The Institut Français des Sciences et Technologies des Transports, de l'Aménagement et des Réseaux (IFSTTAR)*

Created in 2011 from the merging of INRETS (*Institut National de Recherche sur les Transports et leur Sécurité*) and LCPC (*Laboratoire Central des Ponts et Chaussées*), IFSTTAR is a public scientific institution under the supervision of the French Ministry of Ecology, Sustainable Development and Energy and the French Ministry of Higher Education and Research. "*IFSTTAR conducts finalized research and expertise in the fields of transport, infrastructure, natural hazards and the city, in order to improve the living conditions of our citizens and, more generally, to promote the sustainable development of our societies*". The SPLOTT laboratory (*Systèmes Productifs, Logistique, Organisation des Transports, et Travail*), founded in 2005, is attached to IFSTTAR.

http://www.ifsttar.fr/

1

Operators in Logistics

1.1. Sheet 5: Transport and Logistics Jobs

Key point

The nomenclatures of transport and logistics jobs are relatively numerous. They show a wide diversity of jobs. Some highlight jobs in both freight transport and passenger transport, such as Studyrama's nomenclature, while others are only involved in freight transport (see APEC – *Agence pour l'Emploi des Cadres* – Executive Employment Agency). The study conducted by ASLOG and the Michael Page recruitment agency lists 33 types of jobs to date that are encountered in transport and logistics.

1.1.1. *Families of jobs in transport and logistics*

Studyrama[1] presents four families of jobs in transport and logistics:

– transport: ambulance driver, postman, taxi driver, cash courier, delivery man, coach driver, bus driver, metro or tramway driver, train driver, truck driver, movers, merchant navy seaman, naval officer, barge or sailor, captain, first mate, executive driver or master driver;

– logistics: freighter, routing agent, transit agent, logistics agent, professional packager, warehouse or distribution agent, warehouse manager, operations manager, order picker, station manager, safety advisor, rail traffic regulator, freight agency manager and customs declarer;

1 http://www.studyrama.com/formations/fiches-metiers/transport-logistique/.

– sales: commercial agent in railways, commercial agent in air transport, commercial agent in public transport, controller of public transport, industrial vehicle rental company, private car rental company and business manager;

– technical: maintenance agent, handling agent, wire-guided lift operator, chief mechanic and deck mate, railway maintenance operator and training manager.

APEC (*Agence pour l'Emploi des Cadres* – Executive Employment Agency) lists a total of 22 job profiles[2], which are classified into four families:

– Family 1: "flow design" with six job profiles, namely logistics analyst, logistics consultant, supply chain manager, shipping receipt manager, operations manager and platform manager;

– Family 2: "flow management" with seven job profiles, i.e. supplier, product flow manager, warehouse manager, dispatch manager, stock management manager, preparation/receipt manager and technical maintenance manager;

– Family 3: "commercial negotiation" with seven job profiles that are charterer, agent (maritime, air), agency manager, transit manager, line manager, transport purchasing manager and fleet manager;

– Family 4: "rules, standards and contracts" with two profiles, i.e. customs declared responsible for contracts.

1.1.2. *Key Transport and Logistics Functions according to ASLOG and Recruitment Firm, Michael Page*

On March 31, 2015, the study of functions (and remuneration) of logistics and supply chain jobs, co-constructed by the recruitment agency Michael Page and ASLOG (*Association Française de la Supply Chain et de la Logistique* – French Supply Chain and Logistics Association), was published. The general finding of this study highlights the following points:

– *"an increasingly strategic sector of the future for companies with a branch that has more than 1.6 million employees*;

2 https://recruteurs.apec.fr/resource/mediatec/domain1/media27/15198-hqm8c6dq39u.pdf.

– a dynamic job market looking for experienced, internationally oriented profiles: a 22% increase in the hiring rate between March 2014 and February 2015 according to APEC"[3].

ASLOG and the Michael Page recruitment agency conducted a study on the basis of 80,000 profiles of candidates sought in transport and logistics. A total of 33 key functions in three different sectors are listed in Table 1.1.

INDUSTRY
1. Supply Chain Manager: "He/she leads the supply chain and is the interface between the sales department, production, financial services, purchasing and logistics, in order to take into account the requirements of these various departments and optimize the global organization".
2. Logistics Manager: "He/she covers all component and finished products stores, as well as all intra-and inter-site flows. [...] It is differentiated from the supply chain manager by a smaller margin, centered on the management of resources and operations".
3. Supply Chain Coordinator: "He/she covers the entire activity of a factory within a large group or the entire supply chain in PME/PMI. [...] He/she is in charge of all the logistics flows following the evaluation and planning of demand".
4. Logistics Coordinator: "He/she covers all operations and factory resources within a group, or across the entire company in a PME/PMI environment".
5. Head of S&OP (*Sales & Operations Planning*): "He/she works on the development of the Industrial and Commercial Plan (ICP) through a monthly collaborative and decision-making process".
6. Sales Forecasting Manager: "He/she leads meetings to collect and update forecasts [...] and manages activity reporting through key performance indicators".
7. Planning Manager: "He/she provides a global anticipation role, in order to drive supply chain processes".
8. Procurement Manager: "He/she guarantees the availability of components, sub-assemblies and/or sub-contracted finished products at the site, in accordance with the contractual purchasing conditions, while controlling the lowest possible level of stock".
9. Transport Manager: "He/she organizes transport, and the geographical and functional areas assigned to him/her".
10. Logistics Supervisor: "He/she acts as a facilitator capable of optimizing the interface between the logistics/transport, commercial, sales administration teams and logistics/transport partners, in order to improve the quality/cost/delay triptych".
11. Buyer/Transport: "He/she is in charge of purchasing transport services (road, air, sea, inland waterway, etc.) for the company".
12. Sales and Customer Service Supervisor: "He/she acts as the interface between sales, production and logistics".
13. Logistics Methods Engineer: "He/she mainly works in project mode [...] in engineering and re-engineering missions".
14. Information Systems/Supply Chain Project Manager: "He/she identifies the issues, conducts scoping studies and leads projects to implement the supply chain strategy".

3 Source: http://etudes-presse.michaelpage.fr/etudes/etude-2015-de-fonctions-et-remunerations-des-metiers-de-la-logistique-et-de-la-supply-chain-2/.

DISTRIBUTION
15. Director of Supply Chain Operations: "He/she leads the teams of logistics, transportation, sales forecasting, inventory and supply management and sales administration, and even purchasing".
16. Logistics Director: "He/she is the director of the entire logistics and transportation business".
17. Director – Logistics Platform Manager: "He/she is in charge of directing a platform at the operational, administrative, financial and social levels".
18. Operations Manager: "He/she is in charge of managing and supervising the logistics operations on site and coordinating the warehouse's activities".
19. Procurement Manager: "He/she optimizes and manages the flow of goods between vendors and stores through warehouses if necessary".
20. Transport Manager: "He/she organizes transportation as best as can be done, within his/her assigned perimeter".
21. Project Manager Organization Methods: "He/she works on performance optimization in supply chain projects".
22. Customer Service Manager: "He/she is the guarantor of customer satisfaction and loyalty".
23. Import/Export Logistics Manager: "He/she is in charge of administrative order management and international freight forwarding".
24. Sales Administration Manager: "He/she manages sales".
SERVICE SECTOR
25. Director of Business/Division: "He/she offers specialized services and solutions that can be transportation, customs, international trade and/or physical or administrative logistics".
26. Regional Director: "He/she is responsible for a geographic region with multiple logistics sites".
27. Site Manager: "He/she is responsible for a site (storage, cross-docking and industrial logistics provided on customer site)".
28. Operations Manager: "He/she organizes and plans the logistics operations assigned to him/her".
29. Transport Director – Agency Manager: "He/she organizes transport operations in the company's catalogue of services or tailor-made to his/her various clients".
30. Transport Operations Manager: "He/she organizes transport operations for a site or geographical area".
31. Engineer in charge of Methods: "He/she is entrusted with the tasks of starting new activities, new warehouses, re-engineering existing files, improving or standardizing the warehouse's logistics processes".
32. Engineering and Deployment Project Manager: "He/she is in charge of engineering/re-engineering, deployment, improvement or standardization of logistics processes".
33. Key Account Operations Manager – Contract Manager: "He/she is the interface of commerce and operations".

Table 1.1. *The 33 key functions in transport and logistics (source: extract from ASLOG and the Michael Page recruitment agency [PAG 15])*

1.2. Sheet 6: The Supply Chain Manager

Key point

The function of the Supply Chain Manager is the "apotheosis" of the logistics and transport function. It corresponds to the strategic competence that affects the internal and external aspects of the company. It demonstrates the ability to set up large-scale projects that combine various fields of expertise, covering logistics, production, finance and human resources.

1.2.1. *Tasks, activities and skills of the Supply Chain Manager*

In 2011, Livolsi carried out an analysis of the tasks, activities and skills assigned to job offers in France under the title of Supply Chain Manager, from the period of 2001 to 2008. Large companies in the industrial sector recruit this type of profile, with 69% of them expecting a dual language proficiency (English required) and 54% of them having a good grasp of the IT tools dedicated to logistics[4]. Table 1.2 lists the main criteria used in the advertisements reviewed[5].

A look at the job offers posted on the Internet at that time tends to confirm these results. Nevertheless, there are two terms that have been "added", namely transport management and tracking. Furthermore, competences such as "dynamism" and "creativity" aside, it is "autonomy", "empowerment" and "interpersonal skills" that are currently favored.

1.2.2. *The Supply Chain Manager and his/her relationship with other company activities*

The profile of the Supply Chain Manager position, as described by Livolsi [LIV 11], shows the cross-disciplinary skills of the Supply Chain Manager, both intra- and inter-organizationally. In order to describe the Supply Chain Manager's relationship with other activities within the

4 % in the 2006–2008 period.

5 The newspapers studied are *La Tribune, Le Monde, Le Figaro, L'Express, Le Point, Logistiques Magazine, Supply Chain Magazine, Stratégie Logistique, L'Usine Nouvelle* and the *Cadreemploi* website.

company, we will rely on the Porter's classic value chain framework [POR 86]. Still widely used to this day, Porter's value chain regards a company as being composed of core activities (which contribute to the design and sale of the product) and backing or support activities (which support core activities).

Period/analyzed offers	2001–2005 Nb. = 86	2006–2008 Nb. = 66
Assigned tasks:		
Optimization of logistics exchanges	70%	45%
Coordination of operations	2%	18%
Reduced logistics costs	2%	15%
Creation, management and change of the supply chain	16%	18%
Management activities:		
Purchasing	40%	24%
Procurement	81%	79%
Production (implementation of sales and production plans)	40%	57%
Distribution	74%	85%
Sales (forecast)	58%	48%
Logistics	84%	82%
Managerial and organizational skills:		
Organization (changes, processes, etc.)	84%	97%
Management	77%	91%
Management tools (projects, dashboards, lean production, management control, problem-solving, etc.)	63%	85%
Dynamism	25%	33%
Creativity	21%	12%

Table 1.2. *Tasks, activities and skills of the Supply Chain Manager (source: Livolsi [LIV 11])*

Core activities

– "*internal logistics. Activities associated with the receipt, storage and allocation of the means of production required for the product, such as handling, warehousing, inventory control, scheduling of transport and referrals to suppliers*". ➔ As defined by Porter, internal logistics is defined as upstream logistics or procurement. In this respect, the Supply Chain Manager will certainly consider the cost aspects, but will above all observe the deadlines and quality in the goods he/she receives, within a logic of not disrupting the production process. The link with the purchasing department is very important here.

– "*production. Activities associated with the transformation of means of production into finished products, such as machine operation, packaging, assembly, equipment maintenance, verification, printing and facility operations"*. ➔ Currently known as production or in-house logistics.

– *"external logistics. Activities associated with the collection, storage and physical distribution of products to customers, such as warehousing of finished products, handling, operation of delivery vehicles, order processing and scheduling"*. ➔ External logistics mainly relates to downstream or distribution logistics.

– *"marketing and sales. Activities associated with the provision of means by which customers can and are encouraged to purchase the product, such as advertising, promotion, sales force, selection of distribution channels, dealer relations and pricing"*. ➔ Here, it is mainly regarding the selection of distribution channels where the Supply Chain Manager will play a role. Thus, in his/her objective of designing warehouses and/or distribution platforms, he/she verifies the most reliable and easy axes and accesses for receiving and delivering the product. The notion of distribution channels is also defined in the relationship with suppliers. Furthermore, it is also possible at this level to define the package shape, the Supply Chain Manager is better able to determine the packaging, therefore avoiding any breakage and/or damage (linked with Research & Development).

– *"services. Activities associated with the provision of services, aiming to increase or maintain product value, such as installation, repair, training, supply of spare parts and adaptation of the product"*. ➔ With regard to Porter's definition, we find all the notions of reverse logistics. In this respect, the Supply Chain Manager will manage everything related to repair, waste management or remanufacturing, the purpose of which is to give a second life to the product.

Support activities

– *"supplies. Supplies refer to the purchasing function of the means of production used in the company's value chain, not to the means of production themselves. In this purchasing function, we consider the purchase of raw materials as well as machinery, temporary labor, entertainment expenses, such as meals and lodging or strategic consultations"*. ➔ In his/her relations with the purchasing department, the Supply Chain Manager addresses "constraints" of time and quality above all, so that his/her production line does not suffer any stoppages. Obtaining discounts, rebates and reductions is more specifically the role of the buyer. On this point, his/her influence can be assessed on the threshold of quantities to be ordered, in order to respond to the presence of emergency stock.

– "<u>technological development</u>. *Technological development involves a range of activities that can be combined into a set of efforts, in order to improve the product and production process".* ➔ Technological development calls for product and process innovation. On these two points, the Supply Chain Manager has an important role to play. He/she is indeed the one who can consolidate in the fulfillment of the product, with regard to the technological and human resources that are at his/her disposal. In terms of process, the Supply Chain Manager must constantly review his/her processes in order to eliminate redundant tasks and bottlenecks.

– "<u>human resources management</u>. *It consists of the activities involved in recruiting, hiring, training, personal development and payment of all categories of personnel".* ➔ In his/her relationship with HRM, the Supply Chain Manager – due to the technical and physical aspects of the jobs associated with transport and logistics – is the most competent in defining job profiles, the necessary training and/or refresher courses and working conditions.

– "<u>the company's infrastructure</u>. *It consists of a number of activities that include general management, planning, finance, accounting, legal, external relations and quality management".* ➔ In his/her relationship with the Executive Management, the Supply Chain Manager decides on the types and standards of investments to be made (building, machinery, etc.). With finance, he/she checks the solvency of short (stock–customer–vendor relationship) and long (investment; leasing) term. With the quality department, all ISO standards must be consolidated with the Supply Chain Manager. His/her action is also crucial in determining (quality) processes.

Box 1.1. *The value chain (source: Porter [POR 86]) and its interpretation of current logistics*

1.3. Sheet 7: Logisticians and Interactions with Other Internal Departments of the Company

Key point

As previously mentioned, logisticians are called upon in order to interact with several internal departments of a company. Indeed, as a flow specialist, his/her role of coordination involves interfaces.

1.3.1. *Daily relationships with other services*

The logistics department is called upon to collaborate with all the internal operators/services present in the company.

Services	Link(s) with logistics
Executive Management	Through its systemic action, both internally and externally, logistics influences short- and long-term decisions. It is therefore often linked to decision-making, such as investments in technical installations, machinery and equipment (balance sheet nomenclature).
Research and Development	Logistics is a player in innovation, both in a technological and organizational sense. Through the design test phases of new products, it verifies the feasibility of production in terms of time, quality and cost (number of materials used). It also gives its opinion on the type of packaging to be used. Logistics acts on product design. Between theory and fulfillment, we have to think about feasibility. The logistician's technical competence is in play here because he/she is in the best position to specify whether the machines, and/or the so-called direct labor force he/she uses, can produce this new product.
Purchases	Although the purchasing department's principle is to obtain maximum discounts from the supplier, raw materials can only be ordered on this criterion alone. The raw material is the first step in the production process. It is therefore important that it is of quality, at the risk of resulting in excessive waste and/or delays due to lower quality. Similarly, the logistician is led to determine a safety stock[6] for certain products and will therefore ask the buyer to take this surplus into account in the order placed. The logistics department is increasingly involved in the technical aspects of drawing up the contract between the company and the supplier. The issue of the number of load breaks[7] is addressed at this stage. Finally, depending on the specificity of the raw material, the logistician is the person who gives the buyer the possible flexibility, in terms of product fragility and the possible waiting period for delivery.

6 A safety stock is an "additional stock" to deal with certain contingencies (difficulty to obtain a material quickly, willingness to avoid a non-sale in case of a bad sales forecast). The presence of a safety stock is subject to the availability of a place in a warehouse, a cost of ownership bearable by the company, and consideration of a possible use-by date for the product.

7 *In the transport sector, a load break is a stage during which goods or passengers carried by a first vehicle are transferred to a second vehicle, either immediately or after a storage period* (source: wikipedia.org).

Marketing/Commercial	There are many relationships between the logistics department and the marketing/sales department. They concern: – the determination of sales forecasts, which have a direct impact on the Industrial and Commercial Planning (ICP) and therefore on production launches; – feedback from customers' expectations on the use/complexity of the product. It should be noted that the customers' expectations are also reflected in the R&D department; – customer dissatisfaction with quality. For example, inappropriate packaging and/or lack of instructions for use; – the importance of holding (over-)inventories of finished products. Indeed, the marketing/sales department is in direct contact with the customer, so it is able to determine the importance of having a rapid response to the customer's order, which is all the more important if the company is located in a very competitive sector; – on the distribution network, the logistics department is particularly vigilant to the distribution network, both in the establishment of factories/warehouses/platforms and their accessibility. Its advice is a strong element to consider and is carried out in coordination with the marketing department.
IT	Information technology is omnipresent in the life of the logistician: when the number of product references is important, when it is necessary to program and schedule orders, in the follow-up of workstations, etc. The logistician must clearly specify his/her information needs in terms of timing/recording, so that input is easy for all operators involved, both internally and externally, and that there are no errors and delays.
Financial	Like any position in a company, the logistics function is accountable to the financial department. It is therefore obliged to control the costs incurred by its service, and beyond, when its action takes place throughout the entire supply chain. Knowledge of the construction of flagship indicators (via dashboards), activity-based accounting, full cost accounting, etc. are prerequisites for the logistics function.
Staff	The jobs in logistics are complex and require a thorough knowledge of the skills to be identified. The Human Resources Department needs the logistician's expertise

	when preparing job descriptions, whether it be when recruiting or when career development offers are made.
Legal	The question of the contract is at the heart of the discussions. In the face of a possible problem, it is he/she who takes note of the judicial decision. For the logistician, the clauses of the contract must be as precise and protective as possible. In addition to the "standard" contract, environmental constraints imposed by decrees are increasingly being added to this day, and the logistician must now include them in his/her work. Furthermore, from the perspective of outsourcing (IT, transport, inventory management, industrial process part), the outsourcing contract must specify both the management of skills transfer and the future of the employees bound by the agreement. As such, the logistician must clearly be aware of his/her leeway for maneuver.
Quality	The relationship between logistics and quality service was recognized early on [WAG 89]. Logistics relies on a large number of tools, resulting from quality improvement such as, for example, "activity flow sheets", which allow us to understand the steps of a process, or "control sheets", in order to visualize productivity improvements.

Table 1.3. *Examples of relationships between logistics and other internal departments of the company*

1.3.2. *The "new" relationships: the ecology/environment service*

The logistician becomes a privileged operator of ecology. He/she acts on various aspects that affect the environment. And in the event that the structure has a department dedicated to ecology, he/she will be a key interlocutor. The logistician's action regarding ecology is established at several stages of the industrial process, such as:

– beforehand, when thinking about the construction of a product and a service, he/she can be a source of proposals on the technical aspect (format and quality of the parts);

– throughout the process, advising on the use of a particular packaging (pallet, protective film, etc.) for reuse;

– on the upstream–downstream and downstream–downstream chain, he/she studies and seeks to manage the number of waste emitted as best (or rather more accurately, the least) as possible. The objective here is to limit the existence of ultimate waste, that is, that which cannot be reprocessed and is landfilled. He/she also acts on remanufacturing issues, namely giving a second life to an already manufactured product. Therefore, the product does not undergo total manufacturing, since we start from an existing product to either fix or improve it [MIC 10, ELK 10]. It also considers the transportation of goods in full containers on the way out and, if possible, when returning.

He/she can also act ecologically on the construction of a building. From the point of view of construction and/or refurbishment, and according to environmental approaches such as BBC (*Bâtiment Basse Consommation* – Low Energy Consumption Building) or HQE® (*Haute Qualité Environnementale* – High Environmental Quality), the expertise of the logistician is crucial, for example, in the smoothing of the ground in order to avoid shaking when forklift drivers pass by with their equipment (impact on the back).

RT 2012 regulation, BBC label and HEQ® approach (source: http://www.connaissancedesenergies.org/batiments-quelles-differences-entre-rt-2012-bbc-et-hqe-111220 - December 20, 2011)

*The "RT 2012" is the French thermal regulation in effect for new buildings, which has been applied since October 28, 2011, to certain tertiary buildings and housing built in the ANRU (*Agence Nationale pour la Rénovation Urbaine *– National Agency for Urban Renewal) zone. It will apply to all new residential buildings starting January 1, 2013. Among other things, this regulation sets a maximum primary energy consumption threshold ("Cepmax") of 50 kWh/m²/year on average.*

*"BBC" or "Low Energy Building" refers to the highest level of performance of the HPE (*Haute Performance Énergétique *– High Energy Performance) label based on thermal regulations. It is currently allocated to buildings that meet the maximum consumption threshold of 50 kWh/m²/year.*

"HQE" or "High Environmental Quality" does not constitute a regulation nor a label. It is a qualitative and voluntary approach that aims to integrate the principles of sustainable development into the built environment.

The HQE approach has taken two standards into account since 2001:

– the Environmental Management System (EMS), a set of procedures relating to project supervision, put in place during construction by the client;

– the Définition Explicite de la Qualité Environnementale *(DEQE) (Explicit Definition of Environmental Quality), a sort of charter that lists requirements, particularly in terms of eco-construction, health and comfort.*

LAW no. 2015-992 of August 17, 2015 on the energy transition for green growth

JORF no. 0189 of August 18, 2015 (source: https://www.legifrance.gouv.fr/affichTexte.do?cidTexte=JORFTEXT000031044385&categorieLien=id)

This law includes several topics and articles on energy transition. Only the titles of this law are listed here.

Title I: Defining common objectives for a successful energy transition, strengthening France's energy independence and economic competitiveness, preserving human health and the environment and combating climate change.

Title II: Better renovate buildings to save energy, cut costs and create jobs.

Title III: Developing clean transport to improve air quality and protect health.

Title IV: Combating waste and promoting the circular economy: from product design to recycling.

Title V: Favoring renewable energies to diversify our energies and develop the resources of our territories.

Title VI: Strengthening nuclear safety and informing citizens.

Title VII: Simplifying and clarifying procedures to increase efficiency and competitiveness.

Title VIII: Empowering citizens, businesses, territories and the State, in order to act together.

Box 1.2.

1.4. Sheet 8: The Supplier–Customer Relationship

Key point

The glossary of standard definitions (ISO 9000:2005)[8] specifies that:

– "*A supplier is an organization or person that provides a product. It includes, for example, the producer, distributor, retailer, merchant, service or information provider. A supplier may be internal or external to the organization*[9]*. And in a contractual situation, the supplier can be called 'contractor';*

– a customer is an organization or person who receives a product. For example, it could be a consumer, customer, end user, retailer, beneficiary or buyer. A client may be internal or external to the organization."

Of these definitions, any relationship is based on a supplier–customer exchange. In the representation of a supply chain, it is therefore necessary to clearly specify which name(s) of operator(s) is/are concerned by the exchanges induced by a relationship, such as the supplier of raw materials, the industry, the distributor, etc. [GAR 03].

1.4.1. *The supplier, the focal company and the customer according to Lambert et al. [LAM 98]*

Mapping relationships between operators in a supply chain is strategic [GAR 03]. In itself, any relationship is built through a supplier–customer relationship. Still widely used today, the representation of the Lambert *et al.* supply chain [LAM 98] is configured around these two operators, while specifying the focal company. The logistics chain is therefore built around the focal company. Relationships are initially displayed from the focal company, locating the place of the supplier and/or customer. Thus, starting from the focal company, Lambert *et al.* [LAM 98] recommended presenting the suppliers – by levels 1, 2, 3 and 4, level 4 being the supplier furthest from the focal company (the one from which the raw material comes) and

---

8 http://www.e-filipe.org/modules/qualite/glossaire.pdf.

9 According to ISO 9000:2005, an organization is a set of facilities and people with responsibilities, powers and relationships. An organization may be a company, corporation, company, enterprise, business, institution, charity, self-employed person, association, or part or combination thereof.

the customers – by levels 1, 2, 3 and 4, level 4 being the customer furthest from the focal company (the one who receives the finished product).

1.4.2. *The supplier – the industry – the distributor – the end customer: the supply chain quartet*

For a better understanding of a supply chain, it is still preferable to specify the name/status of each operator. In fact, when it comes to the supply chain, the "dual supplier–customer relationship" is somewhat represented through a quartet. Thus, interfaces are established via the supplier (of raw materials), the manufacturer (producer/manufacturer), the distributor (retailer) and the final consumer [PAC 00] (see Figure 1.1).

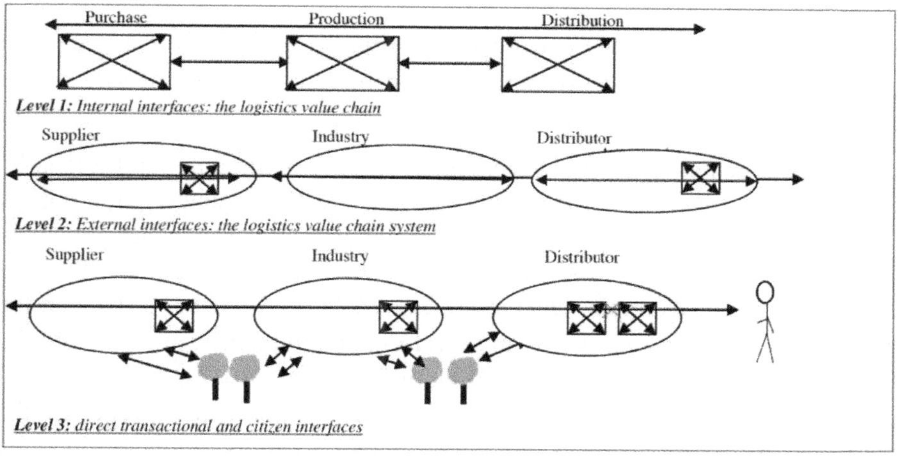

Figure 1.1. *Internal, external, direct and transactional logistics interfaces (source: Paché and Colin [PAC 00])*

Philippe [PHI 97] clarified the roles played by the producer, manufacturer, industry, distributor and retailer:

– the producer is *"the company that designs the product (model, materials), controls its manufacture and sets its sales price (for the first transaction affecting the finished product)"*. The manufacturer is *"the company that owns the machinery necessary for manufacturing"*. As for the industry, it *"ensures the design of the model, the choice of materials, manufacturing, national advertising"*;

– the distributor *"recovers a finished product, highlights it, and if necessary provides specific services (consulting, alterations, after-sales services, credit, local advertising) [...].* The retailer *is an independent merchant, but they never own the merchandise*".

Box 1.3. *Who does what?*

1.4.3. *Other operators in the supply chain*

In a classical architecture where each operator is placed on the same line, Min and Zhou [MIN 02] specified even more operators. Thus, in addition to suppliers, manufacturers, distributors and end customers, their representation identifies retailers and logistics service providers.

Finally, other operators can interfere in the supply chain. Thus, without being a supplier and/or customer, they can influence the act of sale and/or the act of purchase. These include the following:

– governmental and public institutions, both nationally and globally. Standard and regulatory constraints can change the structure of a supply chain. Consumer organizations can also change the way a supply chain operates. Perhaps the most telling example is the constraints of waste reprocessing. For example, Decree no. 2014-928 of August 19, 2014 on the waste management of electrical and electronic equipment[10], arising from the European directive WEEE 2012/19/EU, or Decree no. 2012-22 of January 6, 2012 on the waste management of furniture components[11], requires the recycling of these goods and has forced logistics chains to adapt, either by reprocessing with existing operators in the initial chain or by calling upon other operators (see Sheet 40);

– shareholders who can sometimes impose a particular choice of supply chain configuration (transfer of subsidiaries abroad or, conversely, favor of local logistics);

– professional unions that can, for example, influence a decision to outsource activities.

10 https://www.legifrance.gouv.fr/affichTexte.do?cidTexte=JORFTEXT000029387124&categorieLien=id.

11 https://www.legifrance.gouv.fr/affichTexte.do?cidTexte=JORFTEXT000025114585&categorieLien=id.

1.5. Sheet 9: Operators of Distribution

Key point

Distribution or downstream logistics is the link in the supply chain that has been the subject of early attention from professionals and researchers. It is at this stage that the relationship with the final consumer is closest. As a result, a multitude of jobs appear on this chain.

1.5.1. *Principles of distribution in a supply chain*

Philippe [PHI 97] noted that the overall downstream relationship of a supply chain is grasped through the duality between producer and distributor. Therefore,

> "*The manufacturer ensures the design of the model, choice of materials, manufacturing and national advertising. If there is a real distribution network, it is he who manages it and coordinates the action of its members through a range of contracts under his control: territorial exclusivity, exclusive or selective distribution, linear or twin fares*[12]*, franchising. The distributor, on the other hand, collects a finished product, showcases it, and if necessary, provides specific services (consulting, retouching, after-sales service, credit, local advertising). He is supposed to know the fine details of local demand and possibly inform the producer about it*" [PHI 97, p. 102].

The relationship between producer and distributor therefore seems to be clearly defined. But, as Philippe [PHI 97] pointed out, this is not always the case. In fact, the distributor can intervene in the decision to design the product, in other words, at the producer's premises. Or, some producers are present directly on the distribution market, such as Levi's products.

12 *"A binomial rate is a two-part rate, consisting of a fixed amount (independent of the quantities ordered) plus a unit price"* [PHI 97, p. 102].

1.5.2. *Distribution jobs*

Filser *et al.* [FIL 01] noted that *"the size and complexity of commercial organizations make it essential to coordinate and optimize the exchange of goods and services, but also information flows"*. In this respect, logistics is directly impacted by these exchanges. These exchanges take the form of two types of sales, with or without a store (see Table 1.4). In terms of logistics, sales with and/or without a store[13] thus have consequences for the methods of preparation and delivery of orders. For example, the method of preparing orders made without a store, that is, online, follows two basic models [PAC 08, DUR 10]:

– dedicated on-site preparation in the warehouse. This type of preparation is used when the merchant is mainly active digitally;

– in-store preparation, where there is less electronic activity.

As for the mode(s) of delivery, the location of the warehouses and/or platforms plays a role in view of the place served and the place of origin of the delivered product (see Sheet 34).

1.5.3. *The case for wholesalers*

Also known as wholesale trade, a wholesaler is a business that buys and/or sells goods and/or services to other businesses and professionals. According to the European Communities, wholesale trade is defined as *"any undertaking whose economic activity consists exclusively or principally of reselling goods in its own name either to traders, processors or professional users, including craftsmen or other users. The goods may be resold as is or after processing, treatment or conditioning, as customarily practiced in these professions"* [DUG 00, CAP 13]. We are in a BtoB or Business to Business[14] relationship. Among the traditional wholesalers, the best known are Groupe Métro (food, equipment, supplies, etc.) and Alibaba (products manufactured in China).

13 More commonly used, the expressions for this type of sale are BtoB or Business to Business when the sale is made via a physical store and BtoC or Business to Consumer when the sale is made without a physical store, via computer.

14 An employment training site for trade professionals is dedicated to this activity: http://www.mybtob.fr/.

Store Types[15]	**Specifications**
SALE WITH STORE[16]	
Traditional shop	Assisted sales Specialized assortment
Grocery	Assisted sales Specialized assortment
Large department store: *free entry with fixed-price selling, in other words, displayed in combination with a wide and deep assortment* Examples: Le Printemps and Galeries Lafayette	Assisted sales Non-food, non-specialized broad and deep range of products
Specialized large surface area	Self-service Specialized assortment
Hard discount: *low prices and minimum service*	Self-service Non-specialized assortment
Hypermarket: *department store on the outskirts of urban centers, low prices* Examples: Carrefour and Leclerc	Self-service Non-specialized, food and non-food assortment
Factory Store	Self-service Specialized assortment
Popular Store	Assisted sales Large and small, non-specialized food and non-food assortment
Market	Assisted sales Non-specialized assortment
Small general store in rural area	Assisted sales Non-specialized assortment
Discount store, sales depot	Self-service Specialized assortment
Supermarket Examples: Casino and Intermarché	Self-service Food-dominated non-specialized assortment

15 There are seven main distribution formulas, namely the hypermarket (over 2,500 m²), the supermarket (between 400 and 1,000 m² for the supermarket and 1,000 and 2,500 m² for the large supermarket), the hard discount (between 300 and 1,500 m²), the popular store (on average, 1,500 m²), the convenience store (between 120 and 400 m²) and the specialized food store (on average, 200 m²). (source: http://www.distripedie.com/distripedie/spip.php?article978).

16 Sales with stores are part of the retail trade, that is, the sale of products by unit and/or in small batches with little or no modification. Two types of brands are represented in stores: the one under the manufacturer's brand name and the one that is called MDD, that is, *Marque de Distributeur* (Distributor Brand) which is created and owned by the distributor. Prices offered by MDD's are often lower than the prices of producer brands.

SALE WITHOUT STORE	
Electronic commerce	*Presentation of products on a virtual support, with the possibility to buy online*
Teleshopping Example: M6 boutique	*Television program presenting a product that viewers can order by telephone*
Distribution tour Example: rural grocery truck in rural areas	*Mobile point of sale, serving each dwelling in a geographic area according to a fixed schedule*
Truck sale Example: sale of tooling	*Point of sale that punctually serves a point in a town*
Automatic sale Example entry: Selecta	*A vending machine allows the purchase of some products at a location associated with a strong flow of shoppers*
Sale by catalogue Example: La Redoute	*Periodic production of a catalogue (usually twice a year) from which customers place orders that are delivered to their homes or delivery points*
Sale by meeting Example: Tupperware	*An employee of the company organizes meetings, either at home or with private individuals, to present products and record orders*

Table 1.4. *The different types of sales (source: adapted from Filser et al. [FIL 01])*

It should be noted that logistics providers (see Sheet 10) are increasingly assuming the role of traditional wholesalers [CHA 12, CAP 13].

1.6. Sheet 10: Logistics Service Providers

Key point

Transport and/or inventory management can be a burden for a company. It is binding, as it does not necessarily have the technical means (truck, warehouse/platform formats) and/or human resources (skills). Similarly, these functions may not be considered "strategic" in the core activity of the company. In this case, it will seek to outsource, to make up the work via other operators whose core business is the same. This is the positioning of logistics service providers (PSLs): to position themselves on linkages in the supply chain that are not always considered as value creators. However, as time has passed, the logistics service has evolved from an operational provider to a strategic one; a force for proposals.

1.6.1. *Logistics services: between subcontracting and outsourcing*

The "logistics service provider" is part of the subcontracting business, because it meets a set of specifications previously defined by the client[17]. Subcontracting *"is a mode of organization in which a contractor entrusts a supplier with a task that the supplier must perform according to his instructions"* [HAL 04]. Halley [HAL 04] pointed out that subcontracting differs from outsourcing, which *"refers to the strategy of selling a business and the reasons for this decision"*.

Subcontracting, a punctual contract between principal and subcontractor

Subcontracting is a contract by which a principal entrusts another company – called the "subcontractor" – with the execution of part of its production or the components it needs.

In this context, the subcontractor must therefore carry out work, according to precise specifications, submitted by the principal in order to comply with its own standards.

Subcontracting is usually ad hoc *or temporary and involves a specific mission, so it is a "task-based" contract.*

Outsourcing, a long-term collaboration

Outsourcing consists of entrusting an entire business process or service to another company on a contractual basis, usually on a multi-year basis. Most companies choose to outsource business processes that are not part of their core business.

In general, outsourcing can only concern one service, if an industrial company wishes to outsource one of the elements of its production line to a third party, it will be called subcontracting.

Box 1.4. *Difference between subcontracting and outsourcing (source: extract from Cotraitance.com, 2014[18])*

17 At the supply chain level, subcontracting can also be part of a production process.
18 http://blog.cotraitance.com/externalisation-et-sous-traitance-quelles-sont-les-differences/.

According to Halley [HAL 04], there are three orders of subcontracting, namely:

– capacity subcontracting: secondary, low complexity and low value-added activity;

– specialty subcontracting: support and strategic activity where a strong technical competence of the subcontractor is generally required;

– intelligence subcontracting: secondary, complex and low value-added activity. This includes, for example, delivery and logistics support.

1.6.2. *Levels of expertise in logistics services*

Fulconis *et al.* [FUL 11] pointed out that the emergence of PSLs dates back to the early 1970s in the United Kingdom, with a boom towards the end of the 1980s when specialization appeared in various markets.

Roques and Michrafy [ROQ 03] listed three components of the PSL supply system:

– the core business includes order picking, storage/warehousing, inventory management and transport;

– additional services to the customer which are located in the after-sales service, invoicing for customer accounts and archiving;

– the new business lines which consist of site installation, co-manufacturing, packaging, call center management, co-packing, packing and information technology.

Among the new jobs of the PSL,

– co-manufacturing *is defined as "collaboration between companies, based on the specific skills of each company to jointly produce a final product. These tasks range from product design to manufacturing and assembly. These delayed customization operations are often entrusted to logistics service providers, particularly when they are integrated into flow management operations. It may involve pre-manufacturing as part of the operations carried out during the procurement phase, or post-manufacturing during the distribution of the products"* (source: http://www.cat-logistique.com/vocabulaire.htm);

– co-packing: *refers either to grouping products together in the same packaging in batches for promotional purposes or grouping together items which must be associated for sale. (Example: toy + battery, CD sets, tool sets, etc.)*" (source: http://www.cat-logistique.com/vocabulaire.htm);

– packing: *"it is the preparation of a parcel before shipment. Packing includes packaging and protecting the objects to be shipped"* (source: http://www.linternaute.com/dictionnaire/fr/definition/colisage/). Packing follows the picking operation *"which consists of taking, from their place in stock, the various elements (pallets, parcels or consumer sales units) of an order"* (source: Lexipro [LEX 14]).

Box 1.5. *New jobs of logistics service providers*

The Lexipro published by Logistiques Magazine [LEX 04] categorized logistics service providers into three types of operators:

– *"traditional logistics service providers, who carry out physical logistics operations (transportation and warehousing) and whose management system is limited to the monitoring of the latter, on behalf of the client company;*

– *value-added logistics service providers, who integrate a number of services into the traditional provider's offer, ranging from handling complex handling operations (co-manufacturing, co-packing), to the management of administrative operations (invoicing, orders) and information management (tracking and tracing*[19]*, etc.);*

– *logistics service integrators, who are characterized by the near absence of their own physical means and whose specificity is to integrate the services of different subcontracting companies (transport, warehousing, value-added operations, etc.) and to ensure their coherence and management, by controlling the related information flows".*

The consulting company Arthur Anderson, now called Accenture, offers a classification into levels of expertise in terms of logistics services:

– 1PL for First Party Logistics (Provider), which manages the outsourcing of transport or warehousing;

– 2PL for Second Party Logistics (Provider), which includes the outsourcing of transport and warehousing;

19 See Sheet 36 on traceability.

– 3PL for Third Party Logistics (Provider), logistics service providers who manage the different flows. The services offered are varied, such as transport, warehousing, stock management, return management, etc.;

– LLP for Lead Logistics Provider, which is located halfway between 3PL and 4PL. It provides a service both through its own assets and those of other companies;

– 4PL for Fourth Party Logistics (Provider). The management of logistics flows is only done in an informational way. It acts as a link between the company, the market and other service providers;

– 5PL for Fifth Party Logistics (Provider), which coordinates the activities of various subcontracting companies and designs new logistics IT solutions.

		Logistics CA	Warehouse area	Permanent logistic staff	
1	GEODIS	€670 M	1, 353, 281 m^2	NC	3PL and 4PL for its SCO division: Supply Chain Optimization created in 2009 (2012 data)[20]
2	KUEHNE+NAGEL	€625 M	1, 500, 000 m^2	6 000	4PL in Europe, but 2PL or 3PL depending on the region of the globe (2013 data)[21]
3	NORBERT DENTRESSANGLE	€489 M	2, 000, 000 m^2	5 000	3PL, 4PL and LLP (2013 data)[22]
4	STEF	€465 M	4, 603, 000 m^2	NC	3PL, 4PL
5	ID LOGISTICS	€418 M	1, 910, 000 m^2	NC	3PL

Table 1.5. *Top 5 of the top 100 PLs in France (source:* Issu de Logistiques Magazine, *December 2014 and information on their status in reading the website)*

20 *Prestataires logistiques: du stockage au pilotage de flux, Supply Chain Magazine*, May 2012, no. 64, pp. 42–72 – http://www.supplychainmagazine.fr/TOUTE-INFO/Archives/SCM064/APPELS-OFFRE-64.pdf

21 VON PFUHLSTEIN H., Kuehhe + Nagel: Challenges and adaptation of the operating model, Powerpoint of the Vice President, Head of Corporate Development, November 2013, p. 19, http://www.rolandberger.ch/media/pdf/Roland_Berger_OperatingModelKN20131107.pdf

22 Norbert Dentressangle – *Dossier de presse* 2013, *Des réponses humaines aux enjeux de la supply chain*, 24 p. http://www.norbert-dentressangle.com/var/norbert-dentressangle/storage/original/application/ f4978bf61236e8d39b42053f73c8576a.pdf

1.7. Sheet 11: Operators in the Transport of Goods for Third Parties

Key point

A company can choose to transport its own goods. It can manage its own-account transportation. Or it can outsource this activity. It is therefore in a third-party account management mode and it is part of the logistics services jobs (see Sheet 10). Logistics services include transport and/or storage management. In this way, the management of a warehouse can follow relatively standard rules (see Sheets 17 to 22), as transport management requires different skills and professions.

1.7.1. *Modes of freight transport*

Freight transport can be carried out by six means, alone or combined. There is road, rail, air, sea, inland waterway and postal transport [LEG 08]. ADEME notes that road transport is higher in France than the European average, with 84% compared to 76%, and lower for rail transport: 13% compared to 17% in 2011 (source: ADEME, 2014–2017).

1.7.2. *Operators in freight transport for third parties*

It is possible to classify the operators of transport, either through the typology of professions (carriers, renters of industrial vehicles, transport auxiliaries) or according to the legal acts carried out (haulers, freight forwarders, forwarding agents) [BER 06].

1.7.2.1. *Profession typology: haulers, industrial vehicle rental companies, transport auxiliaries*

The *Nomenclature d'Activité Française* (Nomenclature of French activity) (NAF) codifies three activities in terms of road transport of local goods (602L), long-distance (602M) and relocation (602N). To be able to practice the profession of freight transport, it is necessary to apply for registration in the road transport register (Cerfa no. 14557*03 – Order of December 28, 2011).

Industrial vehicle rental companies offer vehicles with drivers for a fixed period of time, knowing that only the costs associated with the vehicle and the driver are charged to the renter. It should be noted that when this type of rental is long-term, it falls within the typology of own-account rental [BER 06].

Transport auxiliaries are called "freight transport organizers" in the NAF nomenclature. Examples are groupage and courier, express freight, chartering, international transport organization, customs agents and forwarding agents [OBE 87].

1.7.2.2. *Intermediaries: freight forwarders, inland freight brokers*

The freight forwarder is a regulated activity. It is *"also referred to as the freight transport organizer*[23]*, any service provider who organizes and arranges for the transport of goods to be carried out, under his own responsibility and in his own name, in accordance with the modes and means of his choice on the account of a principal"* (Inland Transport Guidance Act, 2003, O.J. no. 301). In itself, the freight forwarder does not transport the goods himself, but uses carriers, considered as his subcontractors. In a similar manner, it is he who chooses the mode(s) of transport. The term "in his own name" means that the carrier employed does not need to know the name of the freight forwarder's customer.

The practice of the freight forwarder profession is codified in the *code de transport* (transport code) and governed by Decree no. 2015-1693 of December 17, 2015[24]. In order to be able to practice his profession, the freight forwarder must hold a professional capacity, as provided for in Article R1422-11 of the transport code: *"In application of article 3°. R. 1422-4, any person wishing to practice the profession of freight forwarder in France, whether he resides in France, in another member state of the European Union or in a State party to the Agreement on the European Economic Area, may apply to the prefect of a region with territorial jurisdiction for recognition of the professional qualifications he has acquired in France or in those States, under the conditions provided for in Articles R. 1422-12 to R. 1422-14-1"*. Professional competence may be

23 The freight forwarder is also referred to as a "road charterer, air or sea freight forwarder, etc.".
24 https://www.legifrance.gouv.fr.

obtained either by written examination, professional experience or the equivalence of a diploma[25].

Decree no. 2013-293 of April 5, 2013, rescinded on May 28, 2014, sets out the clauses of a standard transport commission contract. It includes 16 articles that range from the definition of roles and product specifications to quotations and legal clauses.

Like the freight forwarder, the inland freight broker is a regulated activity. Article L4441-1 of December 1, 2010 of the *code de transport* (transport code) defines it as *"the natural or legal person who is mandated to put clients in touch with public carriers of goods by ship, with a view to concluding a transportation contract between them"*[26].

1.7.2.3. *Implementers: forwarding agent, customs agent, etc.*

Only the freight forwarder or freight broker is a regulated activity, the other activities are covered by private law, as is the case with the forwarding agent. For example, the forwarding agent is considered to be a representative, which means that he carries out a transport operation according to the instructions given to him by his customer. In itself, the forwarding agent's decision-making role is non-existent, and he is only responsible for the faults that are incumbent on him.

Within unregulated activities, but which require approval by the Ministry of Economy and Finance, there are customs agents who carry out customs formalities. Therefore, they do not have the status of freight forwarder.

Other implementers include consignees and handling agents. Consignees are found in the maritime sector and handling agents in the aviation sector. The consignee acts on behalf of the ship-owner or, where appropriate, on behalf of the owner of the goods on the ship. They are also called "line agents" or "general agents" and are governed by the law of January 3, 1969, the decree of June 19, 1969 and Article 51 of the decree of December 31, 1966 on maritime law. Handling agents act on behalf of the transporters.

---

25 http://www.driea.ile-de-france.developpement-durable.gouv.fr/conditions-d-acces-a-la-profession-l-exigence-de-a882.html (June 16, 2017).
26 http://www.legifrance.gouv.fr/affichCodeArticle.do;jsessionid=7C494B148E6EA26442349C7603D75D1F.tpdjo06v_1?idArticle=LEGIARTI000023082626&cidTexte=LEGITEXT000023086525&categorieLien=id&dateTexte=20120710.

1.8. Sheet 12: Operators in Urban Logistics

> **Key point**
>
> Urban logistics is involved in the last level of the supply chain. It takes the delivery of products within cities into account. Several types of operators interfere in this logistics: private operators, consumers, public operators and residents also.

1.8.1. *The classification of Chanut et al. [CHA 12]*

In 2012, Chanut *et al.* categorized the operators of urban logistics into four groups of operators:

– flow inducers, which are manufacturers, distributors, e-commerce operators, etc. There are "classical" or "direct" operators, distributors such as Carrefour, pure players such as Cdiscount and Amazon, and wholesalers. The "second" members of this group include chambers of commerce and trade unions;

– flow regulators include cities, regions and local authorities, as well as local residents' and consumers' associations;

– flow and/or infrastructure managers are associated with logistics service providers and transport operators in the private and public sectors and transport unions, such as AFT-IFTIM;

– flow carriers highlight EDF-type energy suppliers, car manufacturers and leasing companies and research organizations.

1.8.2. *Cerema classification [CER 14]*

Cerema [CER 14] considered three types of operators:

– the first group of operators involves the institutional operators, which are the State and the inter-municipality/municipality. Their roles lie at three levels, in other words, the definition of a *national and local regulatory framework* (e.g. road code and environmental code), the allocation of *financial support* to certain members of the second group of operators, such as shippers or commercial, industrial and tertiary activities, and the application of *regulations* to delivery drivers and freight vehicles;

– the second group of operators includes economic operators. These include commercial, industrial and tertiary activities, the shipper, the logistics service provider, the carrier, subcontractors, delivery drivers and freight vehicles;

– between these two groups, the third group of operators is made up of the urban dweller, the taxpayer and the resident. From its links with the economic operators, it buys and receives goods, and from the links with the institutional operators, it pays taxes and makes demands.

1.8.3. *Urban courier operators [DUC 14]*

Finally, we can also mention the operators of courier services which are operators in urban logistics transport. They are specialized in parcels from 30 to 300 kg. For this reason, Ducret [DUC 14] suggested a categorization of the operators involved in urban courier (see Table 1.6).

1.9. Sheet 13: Reverse Logistics Operators

Key point

Reverse logistics is also referred to as reverse distribution, reversed logistics, return logistics, back logistics, and even retro-logistics. Reverse logistics refers to *"the process of planning, implementing and controlling the efficiency, profitability of raw materials, work-in-progress, finished products and relevant information from the point of use to the point of origin, in order to recover or generate value or to dispose of it in the right way, while ensuring an efficient and environmentally sound use of the resources implemented"* [LAM 03].

Reverse logistics can be handled by several types of operators. You can find the company that designed, produced and sold the product and also manages its return, whether it is in the cycle of repair, redistribution, reconfiguration or destruction. However, these operators can be subcontractors of the company that wants to focus on its core business and outsource this activity. There are also "autonomous" operators that deal with end-of-life products without being attached to a specific company, such as car scrap yards.

Families of operators	*The inheritors*				*Other operators*		*New entrants*				
										Specialized urban couriers	
	Historic national postal operator	Express courier services	Mail-order logistics services	Couriers	Other logistics services	Subcontractors	Network of relay points and instructions	E-commerce operators	Other authorized postal operators	Other specialized urban couriers	Green or neutral couriers
	La Poste Coliposte	DHL; TNT; Fedex; Chronopost; UPS; GLS; France Express; Géodis Calberson	Mondial Relay; Relais Colis	Becycle; Novea; Groupe défense 2000; Les Nouveaux Coursiers; Les Coursiers Verts, etc.	Deret; Green Way; Géodis; Heppner; Astre City, etc.	Dilitrans; Flash; Taxicolis; JEP Transport, etc.	Relais Colis; Point Relais; Kiala; Pickup Services; PackCity; Abricolis	Amazon	Colis Privé	Star's Service; Speed Distribution	Colizen; Green Link; Kangourou Vert; Vert Chez Vous; La Tournée; La Petite Reine; Team Distribution; Les Triporteurs Rennais; Oxipio, etc.
Core business	Delivery of normal and express parcels formation				Majority of heavy freight deliveries	Final delivery without added value	Specialization in urban parcel delivery				
Strategic position with regard to urban parcel distribution	Adapting and diversifying the range of services offered Strengthening their positions or playing a new role in the new segment				Diversifying the offer of B2C services The parcel as a secondary activity but an axis of growth		New services Innovation and sustainability as key words				
Logistics organization	Majority of traditional logistics organizations Subcontracting Some innovative organizations				Mainly classic logistics organization Subcontracting		Innovative logistics organizations				

Table 1.6. *Table of the classification of operators and families of operators in urban parcel distribution (source: Ducret [DUC 14])*

1.9.1. *Different categories of stakeholders in reverse logistics*

In reverse logistics, we find the "standard" operators of a supply chain: producers, distributors, etc. However, when the returned good is an end-of-life good, there are two "new" operators, namely collectors and recyclers. Fleischmann *et al.* [FLE 97] observed six categories of operators in reverse logistics. They are as follows:

– suppliers;

– producers;

– distributors;

– consumers;

– collectors who may be the same operators as distributors, companies specializing in returns and communities;

– recyclers who are industrial producers or specialists.

Monnet [MON 11] subdivided stakeholders in reverse logistics into four groups of operators: third parties, marketers, collectors and operators. Third parties *"design, build and control regulatory tools"*, marketers *"finance the system"* as a whole from the extraction of the material to its destruction, and collectors and operators *"collect, sort, treat and recycle waste in compliance with regulations"*.

As for Carneiro *et al.* [CAR 13], they specified the categorization of Monnet [MON 11] into five stakeholder groups. They are as follows:

– third parties which are involved with government, institutions, associations, federations, lobby groups and the media;

– marketers which include producers, importers, dealers and distributors;

– collectors which include communities/groupings, reinsertion companies, associations, PSLs (logistics service providers) and recycling companies;

– operators which concern processing centers, recovery centers and producers;

– a "new" operator appears in the consumer, where we find households, public and private organizations. It is from them that reverse logistics begins, because they bring (back) the product to the end of its useful life for reprocessing.

1.9.2. *The example of waste treatment operators*

Many decrees are put in place to treat waste and limit the presence of ultimate waste, that is, non-reusable or untreatable waste, which is therefore landfill. Based on the polluter-payer principle, collection policies are part of the so-called "extended producer responsibility" chains (*Ministère de l'environnement, de l'énergie et de la mer* – Ministry of Environment, Energy and the Sea [FRE 16]). There are currently 11 types of waste in these sectors, including waste from household packaging, electrical and electronic equipment (EEE), end-of-life vehicles (ELV), used tires, batteries and accumulators, used textiles, graphic paper, unused medicines, furniture components, household specific release devices and infectious risk care activities produced by patients through self-treatment.

The example of the waste management of electrical and electronic equipment (WEEE) is commonly suggested in the logistics cycle. It is the subject of several directives (Directive 2002/95/EC-ROHS of January 27, 2003 on the restriction of the use of certain hazardous substances in electronic and electrical equipment; Directive 2002/96/CE-WEEE of January 27, 2003 on waste electrical and electronic equipment), decrees (Decree no. 2005-829 of July 20, 2005 on the composition of electrical and electronic equipment and the disposal of waste arising from such equipment (WEEE); Decree no. 2006-646 of May 31, 2006 amending the nomenclature of classified installations) and bylaws[27].

Electrical and electronic equipment comprises five product families which are (1) gray products relating to computer equipment (computers, office automation, telephones); (2) brown audio-visual products (television, hi-fi, video recorder); (3) white products relating to household appliances (small electrical appliances, refrigerators); (4) industrial products (electric cabinets, converters); and (5) manufacturing scrap (circuit breakers, switches). To date, five bylaws dating from 2006 give an approval for organizations whose purpose is to remove and treat this waste or as coordinators (see Table 1.7).

27 http://www.actu-environnement.com/ae/dossiers/DEEE/textes_reglementaires_DEEE.php4.

Bylaws	Status	Name
September 22, 2006 – application of Article 9 of Decree no. 2005-829 of July 20, 2005 (WEEE)	Coordinating body	OCAD3E SAS: Role: "Compensation for the costs of the selective collection of household WEEE supported by local communities. It is the financial link between eco-organizations (Ecologic, Eco-systems, ERP and Récylum[28]) and territorial administrations.
August 9, 2006 – application of Article 14 of Decree no. 2005-829 of July 20, 2005	Body for removal and treatment	Eco-systems – created in July 2005 www.eco-systemes.fr
August 9, 2006 – application of Article 14 of Decree no. 2005-829 of July 20, 2005	Body for removal and treatment	Eco-systems – created in December 2005 www.ecologic-france.com
August 9, 2006 – application of Article 14 of Decree no. 2005-829 of July 20, 2005	Body for removal and treatment	ERP for European Recycling Platform – created in 2002 www.erp-recycling.org
August 9, 2006 – application of Article 14 of Decree no. 2005-829 of July 20, 2005	Body for removal and treatment	Récylum – created in May 26, 2005 www.recylum.com

Table 1.7. *Operators in WEEE management (source: excerpt from http://www.actu-environnement.com/ae/dossiers/DEEE/textes_reglementaires_DEEE. php4)*

28 On June 30, 2017, the eco-organizations Eco-Systems (electrical and electronic devices) and Récylum (used lamps) are grouped together. They became ESR on January 1, 2018 (source: *L'Usine Nouvelle* – http://www.usinenouvelle.com/article/les-eco-organismes-eco-systemes-et-recylum-fusionnent.N560173).

2

Innovation in Logistics

2.1. Sheet 14: Different Types of Innovation

Key point

The Larousse[1] dictionary defines innovation through three aspects: *"1. Introduction into the production and/or sale process of a new product, equipment or procedure; 2. The whole process that takes place from the origin of an idea to its materialization (product launch), including market research, prototype development and early stages of production; 3. Influencing process that leads to social change and the effect of which involves rejecting existing social norms and proposing new ones"*. In an organization or chain of organizations, innovation can be both tangible (technological) and intangible (social and non-technological).

2.1.1. *Social innovation versus technological innovation*

Dandurand [DAN 05] considered two types of innovation in an organization. First, there is a *social innovation* that is seen as *"any new approach, practice, or intervention, or any new product developed to improve a situation or solve a social problem, having found its way into institutions, organizations or communities"*. Social innovation is *fragmented* into organizational, educational, media, urban, (etc.) innovations and considers

1 http://www.larousse.fr/dictionnaires/francais/innovation/43196

that *"the imperatives of a social, psychosocial, organizational, economic, political, (etc.) nature overlap"* in technological innovation. It follows that:

– social innovation is akin to *"citizen initiatives and, upstream or downstream, the impacts of research in social and human sciences, and even in art and literature"*;

– technological innovation is articulated in *"Industrial Research and Development or academic research in the field of health, natural sciences and engineering"*.

2.1.2. *Technological innovation versus non-technological innovation*

The Oslo Manual [OSL 05a, OSL 05b] clarified the concept of technological innovation and, by inversion, non-technological innovation. In the 2005a reference manual, technological innovation is subdivided into technological product innovation and technological process innovation (p. 9, point 24):

– *"Technological product innovation means the development/marketing of a more efficient product, with the aim of providing the consumer with objectively new or improved services. Technological process innovation refers to the development/adoption of new or significantly improved production or distribution methods. It may involve changes affecting – separately or simultaneously – equipment, human resources and working methods"*.

The same manual specifies the concept of non-technological innovation on page 97 and points 437, 438 and 439:

– Point 437: *"In its simplest form, non-technological innovation covers all innovation activities which do not fall within the scope of technological innovation, that is, which are not linked to the launch of a new or substantially modified product or service from a technological point of view, or to the use of a new or substantially modified technological process"*.

– Point 438: *"The main types of non-technological innovation are likely to be purely organizational and administrative innovations, which are not taken into account in surveys on technological innovation, except when they are part of a technological innovation project"*.

– Point 439: *"According to the survey conducted by the Australian Bureau of Statistics (ABS) in 1994, the main types of organizational and administrative innovation are as follows:*

- adoption of advanced management techniques (e.g. total quality management and total quality service);

- significant modification of organizational structures;

- the adoption of entirely new strategic orientations or a significant change in the company's strategic orientations".

The Oslo Manual [OSL 05b] provides examples of innovation in Appendix B (from page 169 onwards). A few excerpts are given in Table 2.1.

TECHNOLOGICAL INNOVATION	
Product innovation	Definition (extract): *"introduction of a new or substantially improved good or service"*. Not considered product innovations: – *"minor modifications or improvements,* – *common improvements,* – *regular seasonal changes (such as clothing lines)"*. Examples of product innovation (extracts). For goods: – *"closure systems for clothing; food products with new functional characteristics (margarine to reduce blood cholesterol levels, etc.); etc."* – *"new services that improve consumer access to certain goods or services, such as the provision and return of rental cars to home; introduction of smart cards and multi-purpose plasticized cards, etc."*
Process innovation	Definition (extract): *"implementation of a new or substantially improved production or distribution method"*. Not considered process innovations: – *"minor changes or improvements,* – *an increase in the means of production or service, by adding manufacturing or logistics systems which are very similar to those already in use"*. Examples of process innovation (extracts). For production: – *"laser cutting tools; computerized equipment for production quality control; etc."* For performance and operations: – *"GPS tracking systems for transport equipment; new software tools to improve supply flows; etc."*

Marketing innovation	Definition: *"implementation of a new marketing method involving significant changes in the design or packaging, placement, promotion or pricing of a product".* Not considered marketing innovations: – *"changes in product design or packaging, product placement, product promotion or pricing that are based on marketing methods that have already been used by the company,* – *seasonal, regular or routine changes in marketing instruments,* – *the use of marketing methods already applied to target a new geographic market or new market segment (e.g. socio-demographic group of customers)".* Examples of marketing innovation (extracts). For design and packaging: – *"implementation of a significant change in the design of a range of furniture, to give it a new look and make it more attractive; etc."* For investment (distribution circuits): – *"introduction of authorization to operate a product; etc."* For pricing: – *"first use of a method of price modulation of a good or service according to demand; etc."* For promotion: – *"first use of trademark; etc."*
NON-TECHNOLOGICAL INNOVATION	
Organizational innovation	Definition (extract): *"implementation of a new organizational method in the practices, organization of the workplace or external relations of a company".* Not considered organizational innovations: – *"changes in business practices, workplace organization or external relations which are based on organizational methods already in use by the company,* – *changes in management strategy, unless accompanied by the introduction of a new organizational method,* – *mergers or acquisitions of other companies".* Examples of process innovation (extracts). For company practices: – *"introduction of an integrated monitoring system for a company's activities (production, finance, strategy, marketing); etc."* For the organization of the workplace: – *"first creation of formal or informal working teams, to improve access and sharing of knowledge from different services such as marketing, research and production; etc."* For external relations: – *"first resort to outsourcing of research or production; etc."*

Table 2.1. *Examples of innovations (source: extracts from the Oslo Manual [OSL 05b])*

2.2. Sheet 15: The Concept of Eco-innovation and Its Application in Logistics

Key point

Foulquier [FOU 12] carried out the historical analysis of the prefix "eco". First, the prefix "eco" is associated with the term "ecology", a term created in 1866. But its origin is older. It comes from the Greek word "*oikos*" which means "habitat" and "*logos*" for "study": *"ecology studies habitat"*. It was in 1866 that the term "ecology" was first used by Ernst Haeckel with the meaning of *"the science of the relationship between the organism and the environment, encompassing in its broadest sense all conditions of existence"*. As it stands, the prefix "eco" extends its relationship between man and nature to the relationship between man and man: *"The question of the relationship between man and nature has been debated for a long time. What is certain, and relatively new, is that the word ecology has spread to include the relations of men among themselves. In many ways, we've gone from a natural science, which remains so, to a social science and political discourse"*. The prefix thus changes from an ecological nature to an ecological and social nature.

2.2.1. *The use of the prefix "eco" in industry*

The association of the prefix "eco" and the industrial world highlights the following terms (source: *Journal Officiel* from February 4, 2010):

– Eco-industry: *"industry that offers products or services aimed at improving or protecting the environment"*.

– Eco-technology: *"a set of industrial processes designed to prevent or reduce the negative environmental effects of products at each stage of their life cycle, as well as human activities"*.

– Eco-certification: *"procedure which guarantees that a product or manufacturing process takes into account, according to a corresponding specification, the protection of the environment"*.

– Eco-design: *"the design of a product, good or service that takes its negative effects on the environment into account, in order to reduce them throughout its life cycle, while striving to preserve its qualities and performance"*.

2.2.2. The "eco" and innovation relationship

With regards to the "eco" and innovation link, Mathieu *et al.* [MAT 10] identified a body of work on the notion of innovation and argued that there are two forms of eco-innovation. Through eco-innovation, these authors considered *"environmentally friendly and socially beneficial innovations"* at the interface between the three pillars of sustainable development. They proposed two types of eco-innovation:

– technological eco-innovation represents *"a set of techniques, processes and products that make it possible to eliminate or reduce nuisances, pollution, the use of natural resources, raw materials and energy"*;

– organizational eco-innovation which encompasses *"any new form of organization, a new definition of roles, a new rule or procedure, a new decision-making or human resources management modality, a new mode of communication or management practice within organizations, which contributes directly or indirectly to the improvement of the state of the environment in the broadest sense, including social and societal aspects"*.

2.2.3. The "eco" and logistics relationship

Drawing inspiration from the work of Mathieu *et al.* [MAT 10], Fernandes and Van Steenberger [FER 13] listed five forms of eco-innovation associated with logistics. Eco-sourcing, eco-logistics, reverse logistics, eco-manufacturing and eco-design can all be found.

Reverse logistics is dealt with in Sheet 35, section 6.3.5. The following definitions are used for the other points:

– Eco-sourcing. This *"includes all supplies made from suppliers, subcontractors, service providers, integrating environmental criteria. These criteria can be applied to all phases of a procurement process, but it is mainly during the definition of requirements that these criteria are applied. Upstream of the procurement process, this involves defining material/product requirements and identifying suppliers. The buyer therefore includes environmental criteria in its specifications in addition to technical, functional and pricing specifications. Downstream, this involves monitoring and evaluating suppliers. This is done in the form of audits and performance indicators in order to monitor compliance with requirements, identify non-compliances/dysfunctions and implement an action plan. In practice, eco-sourcing is more efficient when suppliers are involved in the process on a durable basis, which translates into the co-development and co-design of new products. This generally improves each other's environmental performance and fosters innovation"* [CHA 11].

– Eco-logistics. Also known as green logistics, eco-logistics consists of making *"efforts to measure and minimize the environmental impact of logistics activities"* [ROG 01].

– Eco-manufacturing. Newman and Hanna [NEW 96] are the precursors for implementing environmentally-friendly production through the use of less energy. The evolution so far links innovation and production [OEC 09] through the expression of sustainable manufacturing: *"The connection between sustainable production and eco-innovation underlines the important role of innovation in the traditional transformation of production processes for a more sustainable paradigm"* [KIN 13].

– Eco-design. *"Eco-design concepts were introduced internationally in 2002, with the publication of ISO/TR 14062. It states that eco-design aims to 'integrate environmental concepts into product design and development'. It may be recalled that France defined the first documentation booklet on the subject in 1998 in the now repealed standard, called FDX 30-310. The ADEME contributed strongly to this, then represented France at the international level and acted as secretary for the drafting of the ISO 14062 standard. The French operators were able to rally themselves in order to present the concepts and it was then important to take action. The ADEME has since supported various initiatives. Subsequently, in 2009, Directive 2009/125/EC, laying down a framework for the setting of eco-design requirements for energy-related products, defines eco-design as 'the integration of environmental characteristics into the design of the product with a view to improving the environmental performance of the product throughout its life cycle'"*[2].

2 http://www.ademe.fr/entreprises-monde-agricole/organiser-demarche-environnementale/dossier/eco-concevoir-produits/enjeux-lecoconception-benefices-lentreprise-leconomie-lenvironnement (updated on 18/08/2017).

2.3. Sheet 16: Sustainable Logistics Innovation

Key point

"Sustainable innovation brings together all the innovative initiatives of an organization seeking to combine environmental and social issues with those of its activity. As we understand it, sustainable innovation is not limited to technical innovations in the field of the environment (like cleantech or smart grids). It certainly includes a reflection on the environmental, but also social and economic aspects of innovation. Likewise, sustainable innovation cannot be understood through the sole technical spectrum of the company's R&D experts, as it is likely to mobilize and impact all internal functions (marketing, design, human resources, supply chain, etc.) and the company's external stakeholders" [ACO 13].

2.3.1. *The concept of sustainable innovation*

For Acosta *et al.* [ACO 13], five business models adhere to sustainable innovation approaches:

1) The social and cohesive economy and Bottom of the Pyramid (BOP)[3] approaches.

The concept of the social and cohesive-based economy (SSE) refers to a group of enterprises organized in the form of cooperatives, commons, associations or foundations, whose internal functioning and activities are based on a principle of cohesive and social utility. These companies adopt democratic and participatory management methods. They strictly control the use of the profits they make: individual profit is prohibited and the results are reinvested. Their financial resources are generally partly public. They benefit from a legal framework strengthened by Law no. 2014-856 of July 31, 2014 on the social and cohesive economy.

Box 2.1. *The social and cohesive economy (source: extract from https://www.economie.gouv.fr/cedef/economie-sociale-et-solidaire)*

3 Bottom of the Pyramid (BOP) approaches focus on the poorest groups of people. In 2011, *Les Echos* specified that there were "four billion human beings living on less than $7 a day" (source: http://archives.lesechos.fr/archives/cercle/2011/10/25/cercle_39233.htm).

2) The economy of functionality. This environment-focused approach consists of offering a good or service according to its use. Van Niel [VAN 14] categorized the economy of functionality according to three income models, namely (1) collective consumption, where we find short-term rental, car-sharing, laundromats, etc., (2) sales of functions used in conjunction with leasing[4] and functional sales, and (3) contracting of the result, relating to energy performance contracting, facilities management[5], supply at lower cost and integrated biological protection.

3) The economy of reduction. The objective here is to maintain or increase the level of service, while reducing resource consumption.

4) The collaborative economy adheres to the concepts of co-production, co-location and co-consumption of goods and services.

5) Finally, the circular economy aims to neutralize the environmental impact of goods and services. According to the ADEME [ADE 13], *"the circular economy can be defined as an economic system of exchange and production which, at all stages of the life cycle of products (goods and services), aims to increase the efficiency of resource used and reduce the impact on the environment, while developing the well-being of individuals. The circular economy should aim to globally and drastically reduce the waste of resources in order to decouple resource consumption from GDP growth, while at the same time reducing environmental impacts and increasing well-being. It is about doing more and better with less"*. Also for the ADEME, the circular economy includes sustainable procurement, eco-design, industrial and territorial ecology, economy of functionality, responsible consumption, longer life, recycling and recovery of waste.

In addition, according to Acosta *et al.* [ACO 13], four strategies can be applied in order to ensure the sharing of sustainable innovation: (1) seeking the simplification of innovation; (2) managing a portfolio of projects; (3) creating new brands/divisions; and (4) pursuing an external acquisition policy.

4 Leasing is the renting of an asset that generally corresponds to its economic life, with or without an option to purchase at the end of the contract. According to [VAN 14], if the company chooses to buy, leasing is not a functional economy, but a subcontracting operation.

5 *"Activities which consist of taking over support activities on behalf of a company which delegates them, in their entirety (example: operation and support of the logistics system)"* (source: http://www.cat-logistique.com/vocabulaire.htm).

2.3.2. Sustainable logistics innovation: examples of award-winning companies

In 2015, the professional magazine *Stratégies Logistique*[6] launched a sustainable innovation award in the transport and logistics sector. For this first edition, six companies were rewarded:

– The Urban Logistics Prize: *"The Urban Logistics Prize was awarded to the Drivoo collaborative delivery platform, which connects private individuals (Drivers) making deliveries on their daily journeys. The rewarded service is aimed at local retailers (restaurants, wine shops, florists, etc.), and also at supermarket chains and more generally at all physical sales outlets located in urban areas and requiring short distance deliveries. Created in May 2015, it now has more than 3,000 registered individuals and around 100 businesses regularly using its services".*

– The Eco-design award: *"The CHEP Group has been awarded for its new-generation pallet. CHEP, a company specializing in pallet management and rental, has developed a new eco-designed pallet that combines supply chain and marketing innovation at the heart of the circular economy. Its innovations (reusable as part of the management-lease model, 100% made from recycled materials, fully recyclable, patented notches, light weight, storage space saving...) earned it the Eco-design award".*

– The Infrastructure Prize: *"The SOGARIS' Créteil 3 project, which consisted of choosing not to demolish, but to permanently renovate a 35-year-old logistics platform, already rewarded with the SIMI 2015 Grand Prix, has itself been awarded the Infrastructure Prize".*

– The Carrier/Service Provider Prize: *"TK Blue, an extra-financial Transport rating agency organized as a collaborative platform, received the Carrier/Service Provider Prizes. The agency measures and rates the CSR behavior of companies and values their eco-responsible commitments. The basic idea is to create a quality reference system that is recognized and shared by all the operators in the transport chain, while at the same time, enable them to meet all their new regulatory and legal obligations".*

– The Loaders Prize: *"The Société Coopérative d'Approvisionnement ALSACE 'SCAPALSACE'* (ALSACE Cooperative Supply Company), *which now manages the purchasing, supply, warehousing, transport and all*

6 http://lefret.over-blog.com/2015/12/prix-strategies-logistique-de-l-innovation-durable-les-laureats.html

distribution logistics for E. Leclerc in the Far East was awarded for the 'Premier Entrepôt Robotisé E. Leclerc' (First Robotized E. Leclerc Warehouse) *in France, which combines technological innovation, opportunity in terms of human resources and environmental procedure (ground floor space divided into 2)".*

– The Grand Prix: *"The Grand Prix was awarded to the Ooshop/Deret/LPA alliance for the ELU (urban logistics space) of the Cordeliers (Lyon). Inaugurated in April 2012, it is the first logistics platform in the city center of the Lyon metropolis. Available through LPA, it is now shared by the two operators in a mutual approach. Deret Transporteur thus shares the 320m² of the ELU with Ooshop, as well as the supply of goods to the site, which is carried out in a shared hybrid truck: Deret transports its pallets from its Moins platform, and collects Ooshop's products at its warehouse in Vénissieux".*

3

Warehouse Management

3.1. Sheet 17: Design of a Warehouse

Key point

A warehouse can be found at every level of an intra-organizational supply chain. It is possible to have a warehouse upstream for the stock of raw materials, in-house during the production process for intermediate and/or semi-finished products and downstream for finished products. A warehouse may also be located far from the production site, both for reasons of space shortage in the property and for strategic reasons, in order to be close to the customer. In any event, warehouse management is a central component of business management [ROU 00, ROU 06, GU 07, DEK 07, BAK 09, ROO 09, GU 10, TRE 10, LAU 10].

3.1.1. *The role of a warehouse*

Before disclosing the roles and stakes of a warehouse, it is necessary to remember the difference between a warehouse and a platform, which are often associated, but not so in their daily management. The warehouse is defined as a storage area (more or less substantial, with products of different sizes and life spans) where the storage of products lasts more than 24 hours. Conversely, within a platform, products only transit for shipment to another destination. This transit technique is known as cross-docking.

A warehouse serves as a link between customer demand and production. Indeed, demand fluctuates over time (day – week – seasonal variations) and production wants to be as regular as possible (smooth production), the warehouse keeps the products in storage because of this discrepancy. However, the warehouse can also play an active role in the production process, in that it becomes a place of natural transformation for the product (wine cellars, ripening of bananas, etc.).

Two functions are assigned to the warehouse, that is, storage (see Sheets 19 and 20 for stock types and valuations) and distribution (see Sheet 34 for distribution channels).

3.1.2. *Design of a warehouse*

The design of a warehouse requires a number of considerations and rules in order to ensure its proper operation. Several authors have addressed these:

– Rouwenhorst *et al.* [ROU 00] recommended attention to (1) process (receiving, storing, retrieving and shipping); (2) resources (load units, equipment, computer systems and human resources); and (3) organization (process and policies);

– Baker and Canessa [BAK 09] identified two phases in warehouse design, excluding equipment purchase, construction, installation and project management:

- in the first phase, six elements have to be considered (from most to least important), namely the layout of the warehouse, the definition of storage areas, information management, the type of equipment to be held, the parts management system and the level of personnel;

- in the second phase, 11 steps lay down the architecture: (1) defining the requirements of the system, (2) defining and obtaining data, (3) analyzing data, (4) establishing the unit loads to be used, (5) determining the operating procedures and methods, (6) considering the possible types of equipment and their characteristics, (7) calculating the capacities and quantities of the equipment, (8) the definition of services and ancillary operations, (9) the preparation of possible modifications, (10) evaluation and analysis, and (11) identification of the preferred design.

– Continuing with the work of Goetschalckx and Ashayeri [GOE 89], Gu *et al.* [GU 10] consider five points:

- determining the structure of the entire warehouse (management of material flows, specification of functional departments, relational flows between departments);

- sizing of the warehouse and its departments (size, dimension, space allocation);

- determining of each location within each department (e.g. reconfiguration of lanes and reserve storage area);

- equipment selection (the appropriate level of automation and identification of equipment types by store, transport, order management and sorting) and;

- selection of operational strategies.

These decisions are also subject to a performance evaluation, through the implementation of benchmarking techniques such as the DEA[1] tool, analytical models, lane management analysis models and integrated models that combine time of preparation and quality analysis.

To date, the design of a warehouse must also seek to meet ecological standards, through the application of RT 2012 regulations, the BBC label and the HEQ® approach. This design and construction are also covered by the law of August 17, 2015 on energy transition for green growth (see Sheet 7, Box 7.1 for details of these standards).

3.2. Sheet 18: Layout of a Warehouse

Key point

The layout of a warehouse must be considered in terms of both the physical configuration and the regulations imposed for its operation. While the construction of a warehouse is generally understood as a closed building, it should not be forgotten that some products are stored outdoors and that this external storage must – just as the internal one – be considered and comply with the rules dictated.

1 Data Envelopment Analysis or wrapping method which consists of *a "method of determining the performance of production units, using a graphical representation of a surface area known as the wrapping of individual cases"* (source: http://www.cbanque.com/vocabulaire/methode-enveloppement).

3.2.1. *Areas in a warehouse*

Several zones are present in a warehouse (see Table 3.1). Each zone must be marked horizontally with floor markings and vertically with wall and ceiling signs. Whether we are talking about floors, walls and/or ceilings, they must be adapted to the specifications of the products (e.g. dry, fresh and dangerous products) and the equipment used (e.g. smooth and dense floor). In addition, emergency exits, fire extinguishers and smoke extraction equipment[2] must be provided.

Areas	Description
Reception area	The place between the arrival of the product (dock) and the storage area. It is reserved for inbound flows and constitutes a first control area for goods received (quantity and quality).
Storage area	Area for storage of goods, materials and products. A stockyard is when the storage facility is located outdoors. The storage area includes: – traffic lanes (see below); – shelving adapted to the products stored (product size, packaging, storage boxes, etc.); – a label (name + product code) for each product stored.
Order picking area	Area where the ordered product(s) is/are associated with the order (paper/computer). In the event of a product failure, a partial shipment of the order is generally put in place, with a view to customer satisfaction. The balance of the order is processed later. An order sent is identified to allow tracking (order number, recipient, etc.). This identification is established by a barcode or RFID tag (see Sheet 36).
Shipping area	Area where orders are placed temporarily before placing them in the transport vehicles used.

2 The number and type of fire extinguishers to be placed in a warehouse depends on the surface area and specifications of the products. Article R4227-29 created by decree no. 2008-244 of March 7, 2008 – Art. V specifies that a portable water spray extinguisher with a capacity of six liters must be present for 200 square meters, and on every floor (source: https://www.legifrance.gouv.fr/affichCodeArticle.do?idArticle=LEGIARTI000018532079&cidTexte=LEGITEXT000006072050). For flammable products, Articles R4216-21 to R4216-23 and R4227-22 to R4227-27 are cited. For smoke extraction, Article R4216-13 is considered. It should be noted that for sales areas, these are the articles PE14 – Decree of 22/03/2004 (for smoke extraction), PE26 – Decree of 26/06/2008 (for fire extinguishers) and M38 to M43 – Decree of 25/06/1980, consolidated on 21/02/2017 (for flammable products that we will look at).

Control area(s)	Area dedicated to quantitative controls (matching delivery note, delivery quantity, for example) and qualitative controls (e.g. defects). It is also possible to raise questions regarding standardization in this area. Finally, some warehouses may be affected by physical quarantines (prohibition of access to products) or computer quarantines (waiting for administrative clearance to operate the product) [ROU 08]. Area(s) usually interfaced with receiving, picking and/or shipping areas.
Traffic lanes	Traffic lanes that provide access to the service lanes where goods are deposited. They can be one-way or two-way. They are dimensioned to allow the circulation of agents and handling equipment. Articles R. 4323-50 and R. 4323-52 of the Labor Code, created by Decree no. 2008-244 of March 7, 2008[3], define these.
Spaces and storage devices	They are relative to: – the delimitation of storage areas according to the specific characteristics of the products (date of consumption, dangerousness, volume, etc.); – storage options such as: shelving, mezzanines and lockers.

Table 3.1. *The different areas in a warehouse*

A number of regulations and standards affect the operation and availability of a warehouse. Box 3.1 specifies those related to the traffic lanes.

3 *"Traffic lanes used by mobile work equipment shall be of sufficient gauge, in addition to having a profile enabling them to move safely at the speed specified in the instructions. They are kept free of obstacles"* (source: https://www.legifrance.gouv.fr/affichCodeArticle.do?cidTexte=LEGITEXT000006072050&idArticle=LEGIARTI000018489827&dateTexte=&categorieLien=cid). *"Organizational measures shall be taken to prevent workers on foot from being in the development area of mobile work equipment. Where the presence of workers on foot is nevertheless required for the proper execution of the work, measures shall be taken to prevent them from being injured by such equipment"* (source: https://www.legifrance.gouv.fr/affichCodeArticle.do?idArticle=LEGIARTI000018531415&cidTexte=LEGITEXT000006072050).

Various regulatory texts deal with lanes of traffic:

– Article R. 4323-50 (1) of the Labor Code:

"Traffic lanes used by mobile machinery shall have a gauge and a profile enabling them to move safely at the speed specified in the instructions. They shall be kept free of obstacles".

As an indication, the Order of July 30, 1974 prescribed for permanent traffic lanes of trolleys to be at least a width equal to the width of the trolley and its load increased by one meter, or to the width of two trolleys and their load increased by 1.4 meters, in the case of traffic in both directions.

The width of the service lanes depends on the type of trucks, its dimensions and turning radius, as well as the dimensions of the loads and the working site.

– Article R. 4323-52 (2) of the Labor Code:

"Organizational measures are taken to prevent workers on foot from being in the work equipment development area...".

In order to prevent this risk, this requires the physical separation of trolleys and pedestrian traffic lanes.

Pedestrian walkways must be at least 80 centimeters wide. In the storage area, pedestrian traffic is prohibited, but that of an operator is allowed.

Do not store above traffic lanes unless special precautions are taken: openwork floor (see NF EN ISO 14122-2 August 2001), stabilized load and rear protection.

To ensure that the row ends (feet, lower part of the uprights) are not in danger of being struck, install ladder foot guards or floor rails with sufficient strength with a minimum height of 400 mm.

These guards shall be placed in such a way that they cannot contribute to the overturning of industrial forklifts and shall be adapted to the types of trucks used for servicing these installations. When dimensioning the lane width, these protections must be taken into account.

Also protect intermediate ladder feet that may be struck by trolleys while rolling or maneuvering in lanes. One of the means of protection consists of fixing a protective shoe to the ground in front of them, with a height of at least 400 mm.

Box 3.1. *Traffic lanes (source: http://www.spade-equipements.fr/actualites/systemes-de-rayonnages.html?art=53; accessed February 2017)*

3.2.2. *Types of process or methods of preparation in the warehouse*

The warehouse process or method of preparation may be non-mechanized, mechanized and/or automated. In the case of non-mechanized or manual process, warehouse management is controlled globally by a person. In the mechanized process, the person uses mechanical means to help him/her in his/her action. Finally, in the automated process, human action is not considered, except during the original programming.

Within the chosen process, the method of preparation may be detailed or complete (or in mass) (see Table 3.2).

Type	Method	Specification
The *"pick and pack"*	Detail	The sampled product is stored directly in the package. The warehouse is usually equipped with a sophisticated computerized system and a conveyor belt[4] from the cardboard to the picking points.
The *"pick then pack"*	Detail	The product is transported to a preparation area also known as a table or "station" and then stored in a parcel.
The *"pick to light"*	Detail	Thanks to a computer-assisted control tool, the picker goes to the area where the product is to be picked up. The area where the product is located is equipped with an indicator light and a digital display that allows stock to be reduced and the order to be filled.
The *"pick to belt"*	Complete or in mass	Packages contain large quantities of products. This sampling is associated with the presence of a drainage conveyor.
The *"pick to sort"*	In mass and then in detail	The same reference is taken at the same time, but for several orders. The customer sorting is then carried out.

Table 3.2. *Warehouse preparation methods*

3.2.3. *Equipment in a warehouse*

Roux [ROU 08] listed two types of equipment present in a warehouse: static equipment (pallets or racks, lockers, etc.) and mobile equipment (forklifts, stackers, trolleys, stacker cranes, etc.).

4 Roux [ROU 08] noted that there are generally three types of warehouse conveyors: pallet conveyors, parcel and container conveyors, and overhead conveyors.

3.2.3.1. *Static equipment*

The *ministère du travail, de l'emploi, de la formation professionnelle et du dialogue social* (Ministry of Labor, Employment, Vocational Training and Social Dialogue) states the following: *"The height storage techniques make it possible to optimize the premises as much as possible. The most common storage racks are pallet racking: these are metal shelving made up of ladders and stapled*[5] *rails on posts, which are generally used to store products on pallets*[6]*. The complexity of the dimensioning of shelving requires the skills of the manufacturer in order to guarantee the safety of personnel from the installation design"*[7]. The ED 771 dossier issued by the *Institut National de Recherche et de Sécurité* (National Research and Security Institute) (INRS) and written by Michel Aumas [AUM 08] sets out the regulations/standards, design, construction and use of metal shelving.

It should also be noted that since 2009, several European standards have changed the use and maintenance of pallets, including the:

– standard NF EN 15512, dated June 2009 on "Static steel storage systems – Shelving systems with adjustable pallets – Principles for structural calculations";

– standard NF EN 15635, dated January 2009 on "Steel storage systems – Use and maintenance of storage systems";

– standard NF EN 15878, dated September 2010 on "Static steel storage systems – terms and definitions".

5 A rail is the horizontal part of a pallet rack: "Not more technically than visually, since many manufacturers have adopted very similar color codes: quite often blue for ladders and orange for rails. The explanation is simple: a rail must be light in color, yellow or orange, so that the driver can clearly see it at high or medium height. And historically, orange was often preferred to yellow because the paint didn't contain lead..." (*Supply chain magazine*, [ROG 09]).

6 To save space, the pallets are stacked on top of each other; these are called stackable pallets.

7 http://travail-emploi.gouv.fr/sante-au-travail/prevention-des-risques/autres-dangers-et-risques/article/rayonnage-de-stockage

In addition, this equipment must comply with the "seismic construction rules applicable to buildings" according to the Decree of 22/10/2010 of JORF no. 0248[8]. Similarly, European Eurocodes standards on the design, dimensioning and justification of building structures have an impact on the use of the pallets[9].

3.2.3.2. *Mobile equipment*

Roux [ROU 08] presented a list of 24 mobile devices that can be present in a warehouse. Among this equipment, the following are ascertained as types of apparatus:

– the forklift truck, which is a *"lifting and handling machine for transferring loads in factories and warehouses [...] The first versions were forklift trucks manufactured in 1917 by the English industrialist Eugene Clark in the United States"* (source: Wikipedia). In addition to the devices with truck terminology, there are also pallet trucks (or pallet racks) and stackers (*"A stacker is a motorized lifting equipment, intermediate between a pallet truck and a forklift truck: like the pallet truck, it does not have a seat for the operator, but allows a larger lift (up to 5 m)"*) (source: Wikipedia);

– rotary storage, which includes the stacker crane, the automatic warehouse, the storage tower, the carousel or the rotary store: *"The stacker crane (STC) or storage and retrieval machine (S/R machine) is an automated or non-automated device that allows pallets and parcels to be stored in a rack, often at great heights. It's called Giraffe because its mast can be more than 20 meters high"* (source: Wikipedia).

The overhead crane, which is *"a handling device for lifting and transferring heavy loads"*, straddles the internal and external warehouse (source: Wikipedia).

8 https://www.legifrance.gouv.fr/affichTexte.do?cidTexte=JORFTEXT000022941755&categorieLien=id
9 http://www.eurocodes.fr/

3.3. Sheet 19: Types of Stock

Key point

Although the company is looking for a just-in-time flow policy where there is no stock, the reality is never so simple. In fact, the company is often forced to hold stock because of, for example, a withdrawal from a supplier or an earliest possible delivery to a customer. The question that will then arise for the company is: what level of service should I consider acceptable in my inventory management? By level of service, we mean: what will be the number of orders satisfied, compared to orders processed or received? In itself, the warehousing manager will think about the concepts of safety stock, quantities to order, etc. Among the elements to be considered, the logistician will make a link between the number of references processed and their specific characteristics (size, weight, shelf life, etc.), the size and technical constraints of the warehouse and the criticality of the product (service rate to be applied according to turnover of the product and the turnover recovered by the product).

3.3.1. *What is stock?*

Stock is a set of goods, materials, merchandise, etc. that is owned by a company at a given time. It is the consequence of a flow gap between pushed and pulled flows (see Sheet 2). This means that a company is rarely able to produce immediately as it has to own the materials. In order to save time on delivery, the company also stores intermediate and finished products. The holding of stock thus constitutes a guarantee to ensure the fluidity of the production process. Nevertheless, the stock is associated with a major constraint, which is namely its cost: cost of the warehouse and its management, cost of the management of the products, etc. As a result, one of the logistician's objectives is to maximize techniques in order to limit stock and implement a just-in-time flow policy as much as possible.

When we talk about stock, we mean products or items. The description of a material in a warehouse can have several names (often depending on the information system used). The terms used are SKU (Stock Keeping Unit), Part Number or Product ID.

3.3.2. *The nature of stock*

Several classifications may be used to specify the nature of stock (see Table 3.3):

Type	Description
Input stock	Stock of raw material and/or merchandise.
Output stock	Stock of finished product.
Intermediate stock	Intermediate, semi-finished or work-in-progress stock that appears throughout the production process.
Consignment stock	The stock is with the customer (called, in this frame, the consignee) who only pays the vendor when the stored product is used. It therefore remains the property of the supplier. The customer can return the unsold stock without obligation. Unless there is a contractual clause, in the event of loss, damage or theft of stock, it is the supplier who suffers the cost of the loss.
Advanced stock	The stock is with the supplier, but belongs to the customer. It is assimilated to the consignment stock.
Deported stock	The stock belongs to the supplier, but is placed *"on a site delegated by the manufacturer to a service provider (this site can play the role of a consolidation point)"* [DUR 07].

Table 3.3. *The nature of stock*

3.3.3. *The destination of stock*

Stock can be assigned or common stock. It is called "assigned" when the stock is dedicated to a particular production. It is referred to as "common" when the stored product is used regardless of production.

3.3.4. *The stock and "its quantity"*

Depending on the calculation carried out, stock can have a different title (see Table 3.4).

Tool stock	Stock of parts that do not enter the production process, but are used for the smooth running of the company. It can be assimilated to the minimum stock.
Minimum stock	Minimum quantity in stock that corresponds to the delivery time.
Safety stock	Also known as the buffer stock, it corresponds to a level of stock that is constantly maintained in the warehouse and which limits stock shortages due to contingencies, such as forecasts that do not match demand, longer than expected supply times, etc.
Alert stock	Corresponds to minimum stock + safety stock.
Physical stock	Actual quantities in the warehouse. As a reminder, a company must carry out a physical counting of all its parts once a year.
Actual stock	Physical stock + supplier orders in the course of delivery - customer orders in the course of delivery.
Dormant or dead stock	Stock whose turnover is insignificant or zero. Costly and cumbersome, they are usually destroyed.

Table 3.4. *Different types of stock*

3.3.5. *Stock rotation*

There are two types of stock rotation:

– first, so-called non-zero rotation gives the possibility of reselling the stock in the following period. The product is not subject to a sell-by date[10];

– so-called zero rotation makes it impossible to resell in the following period. In this context, the product has a sell-by date.

10 "*For perishable foods, this is referred to as the "use-by date". This limit is mandatory. It applies to sensitive foods, often rich in water, which are likely to pose a danger to human health after a short period of time. This use-by date is indicated on the packaging by the phrase "To consume until...", followed by the deadline set by law or by the producer. Products such as yoghurts, fresh meat, fish, unsterilized ready meals... generally have a use-by-date. They must be withdrawn from sale once the date has passed. Conversely, some products have a "best-before date" (a term that has replaced the optimal use-by-date since the implementation of Regulation (EU) no. 1169/2011 on consumer information on foodstuffs on December 13, 2014. This is the case, for example, with dry cakes or cans. Once the date has passed, the food presents no danger, but may have lost all or part of its qualities: taste, texture... On the packaging, the best-before-date is indicated by the phrase "To consume preferably before...*"" (source: http://agriculture.gouv.fr/alimentation/comment-conserver-les-aliments-dlc-dluo-quelle-difference, 24/02/2017).

3.3.6. *The support of goods: the pallet*

Talking about stock leads to presenting the pallet. The pallet is a wooden, plastic or steel structure on which the goods are placed when they are moved. The loading of goods on a pallet is called palletization, which makes it possible to (1) facilitate handling operations, (2) easily count goods, and (3) protect and ensure the stability of the goods during transport. Usually, the stability of the goods is achieved through a filming operation, which consists of wrapping the goods with a plastic film. The stability of a pallet is facilitated when the stored goods are of the same reference and therefore have identical packaging. In this case, it is called a homogeneous pallet. Conversely, when products are not identical, the term heterogeneous pallet is used.

"Skids date back to Ancient Egypt and Ancient Mesopotamia, at least as far back as the 1st millennium B.C.

The development of the forklift and the needs of World War II logistics operations led to substantial use of pallets.

References to the early modern pallet types are slim with a string of patents showing parts of the development. The earliest may be a U.S. patent on a skid from 1924 describing Howard T. Hallowell's "Lift Truck Platform". In the late 1930s, pallets became more commonplace with the newer forklift types. George Raymond filed for a patent in 1938 (granted US Patent 2178646 in 1939) for a pallet designed to complement a new pallet jack design; the essential features of both are still in common use today. A 1939 patent from Carl Clark shows a type of pallet with steel stringers. War time developments were often just patented after the war, so there is a patent from Robert Braun on a four-way pallet in 1945, and a patent from Norman Cahners (a U.S. Navy Supply Officer) shows a disposable pallet type in 1949. The complete construction principle of a modern four-way pallet is described by Darling Graeme in 1949."

Box 3.2. *History of the pallet (source: https://en.wikipedia.org/wiki/Pallet)*

Several ISO standards are cited for pallet use. These are listed in the catalogue of standards *55.180.20 – General-purpose pallets* on the ISO standards website[11].

11 https://www.iso.org/fr/ics/55.180.20/x/

Currently, the most commonly used pallet is the "Europe pallet" or "EUR-EPAL"[12] or Euro-pallets. Its standard dimensions are 800 mm wide by 1200 mm long. It is designed to support a weight of 1500 kg. The Europe pallet is also available in half-pallets with dimensions of 800 mm by 600 mm. The NF EN 13698-1 standard of October 2003 specifies the manufacturing of European pallets.

3.4. Sheet 20: Inventory Procedures

Key point

Several inventory procedures are possible, depending on the obligations of the *Code de Commerce et/ou de choix stratégiques* (French Commercial Code and/or strategic choices). A stock inventory consists of the identification, counting and assessment of stock in the company. Article L123-12 of September 21, 2000 of the French Commercial Code states that all companies *"must check, at least once every twelve months, the existence and value of assets and liabilities of the company's estate"*.

3.4.1. *Physical, permanent, intermittent and rotating inventory*

The physical inventory of stock consists of a manual counting of products in the warehouse. Its role is to determine possible discrepancies between the accounting entry and reality. The physical inventory thus highlights omissions, theft, breakage, bad codifications, bad storage, etc.

Several counting methods are used:

– first, an intermittent inventory is established once a year, at the time of financial closing and where it is advisable to stop any movement of goods during this physical inspection of the stocked parts;

– second, there is the permanent inventory, which consists of a continuous management of the stock, thanks to adapted software. This inventory method is particularly applicable for references in limited numbers and where counting is relatively easy;

12 The French national committee Epal© is authorized to build pallets in France according to the current laws in place – http://qualipal.fr/

– finally, the rotating inventory consists of counting certain references throughout the year. In this context, the company applies 20/80 type classification techniques, known as Pareto or A-B-C. The application of this practice is popular because it limits a blockage over several days, in case of intermittent inventory, and also allows a more regular monitoring of accounting.

3.4.2. *The Pareto Principle and the A-B-C classification*

The 20/80 Law or Pareto Principle is named after its designer Vilfredo Pareto (1848–1923), who observed that 20% of the population owned 80% of Italy's wealth. This observation was subsequently applied in several areas, including inventory management. The variant A-B-C developed in 1954 by Joseph Juran (1904–2008) consists of a subdivision into three parts: the third part being considered as the residual part, which is the subject of less interest.

Faced with a large number of stock references, the A-B-C, or even A-B-C-D method, is used to categorize stock and establish a relatively regular monitoring. For example, class A, which corresponds to the most critical category of stock, will be subject to the establishment of a systematic safety stock on each product or a permanent inventory.

The breakdown into categories A-B-C conforms to the following logic:

– Class A: 20% of the references, which represent approximately 80% of the total value of the stock or turnover excluding VAT;

– Class B: 30% of the following references, which represent approximately 15% of the total value of the stock or turnover excluding VAT;

– Class C: the remaining 50% of the remaining references, which represent approximately 5% of the total value of the stock or turnover excluding VAT.

In itself, this division into three classes is not a rule. The company can perfectly multiply the number of categories if it believes that this will lead to better inventory management. In this way, it is advisable to keep Class A as "fixed".

The interpretation of the A-B-C classification is as follows:

– we are thinking in terms of values sold and not quantities sold. For example, for 1,000 products sold at 1 euro and 1 product sold at 1,000 euros, it is the product sold at 1,000 euros that is considered critical;

– for example, if 20% of references for Class A or 30% of references for Class B are stated, it is estimated that Class A must contain *a minima* 20% of the references, and Class B must contain *a minima* 30% of the references. Thus, if the classification highlights a reference with 19.9% and the next one with 20.1%, we include the reference that includes the 20.1% in category A. This rule also applies to category B. Consequently, the balance of references is placed in category C;

– it is the percentage of the reference classes that prevails and not the percentage of the monetary value of the stock. If, for example, the value of the stock attached to Class A equals 50% of the total stock value, this percentage is left at 50% and no attempt is made to reach 80% of the value, as stated in the theoretical rule;

– the unit of value used is the total stock value. In order to ensure a good analysis, it is customary to consider that the unit of value is based on the same method of calculation. If the company opts for a calculation method based on the First-In-First-Out and Weighted Average Unit Cost methods (see Sheet 21), the analysis may be lacking unicity. If this is the case, it is preferable to take the sales price excluding VAT as a unit of value: excluding tax, as the same principle applies in the case of multiple VAT rates (Value-Added Tax);

– a breakdown according to the A-B-C method can be carried out annually or at other intervals. In fact, the seasonal factor must be taken into account when selecting the breakdown period. But beware of the implementation of the number of breakdowns of stock, as the more that is cut in the year, the more there is a risk that the categories will vary.

3.4.3. *Example of calculation of the A-B-C method*

Suppose there are 15 references in stock. In the example discussed, the value unit is the turnover excluding VAT. The classification is based on annual sales.

Reference	Annual demand	Unit selling price excluding VAT	Turnover excluding VAT
A – 1	6200	€5.35	€33,170.00
B – 2	902	€21.50	€19,393.00
C – 3	4	€1813.00	€7,252.00
D – 4	2999	€40.25	€120,709.75
E – 5	1325	€15.20	€20,140.00
F – 6	435	€10.50	€4,567.50
G – 7	1	€3750.00	€3,750.00
H – 8	1324	€11.00	€14,564.00
I – 9	456	€49.00	€22,344.00
J – 10	286	€12.00	€3,432.00
K – 11	15	€143.60	€2,154.00
L – 12	236	€11.00	€2,596.00
M – 13	974	€20.00	€19,480.00
N – 14	159	€82.00	€13,038.00
O – 15	123	€73.00	€8,979.00

Table 3.5. *Stock sheet*

Reference n. = 15	Annual demand	Unit selling price excluding VAT	C.A. total excluding VAT	C.A. accumulated	% accumulated C.A.	% references	Classes
D – 4	2999	€40.25	€120,709.75	€120,709.75	40.84%	1 / 15* = 6.67%	**Class A** 3 references (20%) represent 55.22% of C.A.**
A – 1	6200	€5.35	€33,170.00	€153,879.75	52.06%	2 / 15 = 13.33%	
I – 9	456	€49.00	€22,344.00	€163,223.75	55.22%	3 / 15 = 20.00%	
E – 5	1325	€15.20	€20,140.00	€196,363.75	66.44%	4 / 15 = 26.67%	**Class B** 5 references (33.33%)*** represent 33.71% of C.A.
M – 13	974	€20.00	€19,480.00	€215,843.75	73.03%	5 / 15 = 33.33%	
B – 2	902	€21.50	€19,393.00	€235,236.75	79.59%	6 / 15 = 40.00%	
H – 8	1324	€11.00	€14,564.00	€249,800.75	84.52%	7 / 15 = 46.67%	
N – 14	159	€82.00	€13,038.00	€262,838.75	88.93%	8 / 15 = 53.33%	
O – 15	123	€73.00	€8,979.00	€271,817.75	91.96%	9 / 15 = 60.00%	**Class C** 7 references (46.67%) represent 11.07% of C.A.
C – 3	4	€1813.00	€7,252.00	€279,069.75	94.42%	10 / 15 = 66.67%	
F – 6	435	€10.50	€4,567.50	€283,637.25	95.96%	11 / 15 = 73.33%	
G – 7	1	€3,750.00	€3,750.00	€287,387.25	97.23%	12 / 15 = 80.00%	
J – 10	286	€12.00	€3,432.00	€290,819.25	98.39%	13 / 15 = 86.67%	
L – 12	236	€11.00	€2,596.00	€293,415.25	99.27%	14 / 15 = 93.33%	
K – 11	15	€143.60	€2,154.00	€295,569.25	100.00%	15 / 15 = 100.00%	

**: the percentage is determined by making the "place" of the reference on the total number of references, which is 15 references here.*

***: the example shows here that the board of directors does not correspond to 80%; the percentage of references prevails.*

****: 33.33% is determined by subtracting 53.33% - 20.00% (Class A limit).*

Table 3.6. *A-B-C Ranking*

The steps to be taken to implement the A-B-C classification are as follows:

– classify turnover by reference from highest to lowest;

– determine the cumulative percentage of turnover per reference;

– determine the percentage of references. For the first reference, we calculate 1 out of the total of references which is 15 here, and so on. This column is used to classify categories;

– break down the classes in view of the results.

3.5. Sheet 21: The Supply Quantity

Key point

Several models are proposed to deal with the quantity of supply to be ordered, namely the control point model[13] and the calendar management model[14], where the Harris–Wilson formula calculation method is applied, also known as the economic quantity calculation.

3.5.1. *Types of supply*

There are four types of supply that apply depending on the nature of the product and its sales frequency (see Table 3.7).

Frequency	Types	
Variable date/Fixed quantity	Control point method; also referred to as control threshold or replenishment threshold.	Requires a knowledge of: – the average daily consumption, – the lead time (in days): transport + administration + etc.: generally corresponds to the safety stock.
Variable date/Variable quantity	Frequent replenishment.	Need to determine an optimum stock for each product.
Fixed date/Fixed quantity	Calendar maintenance – Frequent fixed replenishment.	The simplest method, but with a risk of: – inflation of the stock, – or out of stock.
Fixed date/Variable quantity	Calendar management – Frequent replenishment.	Need to determine an optimum stock for each product.

Table 3.7. *Types of supply*

13 The control point model consists of triggering the purchase order when alert stock is reached. This method is used for highly seasonal activities.

14 The calendar managed model triggers the order at regular intervals. This method is used for low seasonality activities.

3.5.2. *Calculation methods*

3.5.2.1. *Economical Order Quantity (EQQ) or Harris–Wilson Method*

Introduced by Ford Whitman Harris [HAR 13a, HAR 13b, HAR 15] – Westinghouse engineer, the Economical Order Quantity method (EQC) was popularized by the consultant R. H. Wilson [WIL 27, WIL 34], who made it into an intensive application [ERL 90]. Although often referred to as the "Wilson Method", some authors associate the name Harris with it to emphasize its origin, resulting in the title "Harris–Wilson Method".

The Economic Order Quantity method is mainly valid for regular orders with constant supply rates. It takes place in the case of the control management point (q, s), where q is the quantity to be ordered and s is the control point of the triggered command. This method aims to optimize inventory management by balancing inventory management costs with the cost of placing an order. While this method is widespread, it is nonetheless subject to a number of criticisms, such as the fact that it does not take storage capacity, product obsolescence and, of course, market fluctuations into account. It should also be noted that this calculation formula only applies to one reference each time.

The Economic Order Quantity formula is established as such:

$$\sqrt{}(2.(Cpa.K))/(Pu.i)$$

where *Cpa* is the average annual demand (365 days or 12 months or 52 weeks), *K* is the cost of launching an order[15], *Pu* is the cost of acquiring an order and *i* is the possession or opportunity rate.

EXAMPLE.– The average weekly demand for Product A is 380 units. The unit acquisition cost of an order is 280 €, that of the cost of launching an order of 225 €. The opportunity rate is 5%.

The economic quantity is equal to

$$\sqrt{}(2\,.(380\;units * 52\;weeks)\,.225\;euros)/\;(280\;euros\,.5/100)$$
$$= 796.9585\;or\;797\;units$$

15 The cost of launching an order includes the cost of placing an order + the cost of transport + the cost of receiving it.

When s is reached, the control point will command 797 units of A.

3.5.2.2. *Frequent replenishment*

Frequent replenishment consists of ordering an x quantity when the desired stock level is reached. For example, this desired stock level may correspond to the safety stock. In a policy of frequent replenishment, we generally associate an "optimum stock" not to be exceeded. This optimum stock should make it possible to avoid over-stocking, in addition to taking the storage area of the warehouse into account.

3.5.2.3. *Order according to necessity*

The product is ordered according to the necessity. Here, we are mainly in a pull flow policy, where the product is either quickly available on the market or the product design time is not limited (e.g. luxury product).

3.6. Sheet 22: The Valuation of Stock

Key point

Three methods are available in order to value inventory throughout the process, but only two of them are accepted by the *normes internationales d'information financière* IAS/IFRS (International Financial Reporting Standards), namely the *Coût Unitaire Moyen Pondéré* (CUMP) (Weighted Average Unit Cost or WAUC) and the First-In-First-In-Out (FIFO). The WAUC method is a weighted average of the quantities (and values) used, it is mainly used for non-perishable materials. The FIFO method consists of valuing the inventory at the cost of the value of the first input. This method is generally applied for perishable materials. Finally, the third method – which is not authorized in accounting terms – is the Last-In-First-Out (LIFO) method. This method is used commercially for products whose prices are subject to seasonal variations.

3.6.1. *When to apply stock valuation methods?*

It is possible to apply three methods of stock valuation: the Weighted Average Unit Cost method (at the end of the period or after each input), the First-In-First-Out method and the Last-In-First-Out method, but only the first two methods are authorized for tax purposes.

When a company chooses its method of valuing stock, it is advisable to keep it over several years, because it is the guarantor of a "true and fair view of the assets and liabilities". The company decides when its stock should be valued according to its needs, but it must do so at least once a year when drawing up its balance sheet. For information, the stock valuation is found in the balance sheet, a document that serves as a basis for the company's financial diagnoses.

The choice of methods for valuing stock reflects the following reasons:

– for the Weighted Average Unit Cost method:

- at the end of the period, a low variability in the cost of stock is considered a criterion;

- after each input, the critical criterion of the product is assessed, either because of a high variability in cost or because the product is judged to be a high value-added product.

– for the First-In-First-Out method, perishable products are opted for, by the use-by-date. The oldest products are the ones released the earliest.

3.6.2. *Example of calculation using the Weighted Average Unit Cost method at the end of the period*

The calculation of the Weighted Average Unit Cost at the end of the period is easy to do. We take the sum of all the inputs held in stock over the estimated period, including the initial stock divided by the total quantity of the inputs, and apply the unit cost found to all outputs during the period.

STATEMENT.– The inputs and outputs of Product A are as follows for the month of April 2017:

– Initial stock: 10 units at 60.00 euros each;

– April 5: input slip 4.1 of 25 units at 61 euros each;

– April 6: output slip S04.1 of 10 units;

– April 7: output slip S04.2 of 12 units;

– April 12: input slip 4.2 from 20 units to 63 euros each;

– April 16: output slip S04.3 of 16 units;

– April 21: output slip S04.4 of 4 units;

– April 26: input slip 4.3 from 15 units to 62 euros each;

– April 28: output slip S04.5 of 20 units.

Date	Label	In			Out			Remaining stock		
04/01/17	I.S.	10	€60.00	€600.00				10	€60.00	€600.00
04/05/17	IS4.1	25	€61.00	€1,525.00				35		
04/06/17	OS04.1				10	€61.64	€616.40	25		
04/07/17	OS04.2				12	€61.64	€739.68	13		
04/12/17	IS4.2	20	€63.00	€1,260.00				33		
04/16/17	OS04.3				16	€61.64	€986.24	17		
04/21/17	OS04.4				4	€61.64	€246.56	13		
04/26/17	IS4.3	15	€62.00	€930.00				28		
04/28/17	OS04.5				20	€61.64	€1,232.80	8		
TOTAL		70	€61.64	€4,315.00				8	€61.64	€493.12

Table 3.8. *Example of stock calculation using the WAUC method at the end of the period*

The Weighted Average Unit Cost calculation at the end of the period is equal to (600+1525+1260+930)/(10+25+20+15) or 4315/70 = 61.64285714 rounded to 61.64 euros. Withdrawals from the month's stock and the final stock are valued at 61.64 euros each.

3.6.3. *Example of calculation using the Weighted Average Unit Cost method after each input*

On the same practice statement as in section 3.6.2, the Product A stock sheet, using the WAUC method after each input, is as follows:

Date	Label	In			Out			Remaining stock		
04/01/17	I.S.	10	€60.00	€600.00				10	€60.00	€600.00
04/05/17	IS4.1	25	€61.00	€1,525.00				35	€60.71	€2,124.85
04/06/17	OS04.1				10	€60.71	€617.10	25	€60.71	€1,517.75
04/07/17	OS04.2				12	€60.71	€740.52	13	€60.71	€789.23
04/12/17	IS4.2	20	€63.00	€1,260.00				33	€62.10	€2,049.30
04/16/17	OS04.3				16	€62.10	€999.84	17	€62.10	€1,055.70
04/21/17	OS04.4				4	€62.10	€249.96	13	€62.10	€807.30
04/26/17	IS4.3	15	€62.00	€930.00				28	€60.98	€1,707.44
04/28/17	OS04.5				20	€60.98	€1223.20	8	€60.98	€487.84
TOTAL		70						8	€60.98	€487.84

Table 3.9. *Example of stock calculation using the WAUC method after each input*

WAUC after IS4.1 = (600+1525)/(10+25) = 2125/35 = 60.71428 euros rounded to €60.71. This WAUC of 60.71 € applies to OS04.1 and OS04.2.

WAUC after IS4.2 = (789.23+1260)/(13+20) = 2049.23/33 = 62.0978 euros rounded to €62.10. This WAUC of 62.10 € applies to OS04.3 and OS04.4.

WAUC after IS4.3 = (807.30+900)/(13+15) = 1707.30/28 = 60.975 euros rounded to €60.98. This WAUC of 61.16 € applies to OS04.5.

The final stock amounts to 8 units at 60.98 euros each.

3.6.4. *Example of calculation according to the First-In-First-Out method*

On the same practice statement as in section 3.6.2, the stock record for Product A using the First-In-First-Out method is as follows:

Date	Label	In			Out			Remaining stock		
04/01/17	I.S.	10	€60.00	€600.00				10	€60.00	€600.00
04/05/17	IS4.1	25	€61.00	€1525.00				10	€60.00	€600.00
								25	€61.00	€1525.00
04/06/17	OS04.1				10	€60.00	€600.00	25	€61.00	€1525.00
04/07/17	OS04.2		25 - 12		12	€61.00	€732.00	13	€61.00	€793.00
								13	€61.00	€793.00
04/12/17	IS4.2	20	€63.00	€1260.00				20	€63.00	€1260.00
04/16/17	OS04.3				13	€61.00	€793.00			
	OS04.3				3	€63.00	€189.00	17	€63.00	€1071.00
04/21/17	OS04.4				4	€63.00	€252.00	13	€63.00	€819.00
04/26/17	IS4.3	15	€62.00	€930.00				13	€63.00	€819.00
								15	€62.00	€930.00
04/28/17	OS04.5				13	€63.00	€819.00			
	OS04.5				7	€62.00	€434.00	8	€62.00	€496.00
TOTAL		70						8	€62.00	€496.00

Table 3.10. *Example of inventory calculation using the FIFO method*

In the First-In-First-Out method, the valuation of outgoing stock is carried out on quantities and values in stock, in order to cover the withdrawal quantity. For example, on the output of 16/04/2017 – OS04.3, the quantity is 16 units. Remaining stock shows a remaining stock of

13 units at 61 euros each, and a stock of 20 units at 63 euros each, corresponding to the input on 12/04/2017. To value the output of 16 units, we are forced to make a double valuation: a first one which balances the "old" stock of 13 units at 61 euros each, and a second one of 3 units at 63 euros each on the input of 12/04/2017.

The final stock amounts to 8 units at 62 euros each.

4

Upstream and Supply Logistics

4.1. Sheet 23: The Relationship Between the Purchasing and Logistics Department

Key point

The relationship between Purchasing and Logistics is important. Their combined interests give rise to the fluidity and robustness of the supply chain. If it is acknowledged that the Purchasing department has a policy of cost/quality ratio of the purchased product, logistics systematically adds the deadline. It is therefore essential when ordering a product to combine the two departments, so that everyone can find the strongest and most reliable ratio in the overall global performance of the supply chain. This relationship is being increasingly studied in the academic literature [CAL 06, DUR 07, BRU 08].

4.1.1. *The buyer's tasks*

The AFNOR 50-128 standard [AFN 90] specified that a buyer is an operator capable of *"fostering a creative spirit of competition among suppliers, promoting suppliers' proposals and their suggestions for reducing costs and/or improving performance and informing other functions of changes affecting the environment"*. The purchase concerns different types of products such as raw materials, components, production equipment, packaging, subcontracts, advertising space, etc. A large majority of the purchase concerns the industrial process and thus requires concerted links with the logistician.

The buyer's tasks can be summed up in five points, ranging from the most strategic to the most operational, and it can be noted that links with the logistics activity are omnipresent:

– definition of the purchasing strategy, in conjunction with the engineering department to find the best suppliers. Beyond the cost/quality/time ratio, the search for partnerships based on innovation is privileged. Similarly, attention is paid to all environmental and social aspects, in order to promote a relationship based on a sustainable development approach;

– definition of the purchasing policy through competition, but particularly cooperation with suppliers where operational partnerships are specified in the Just-in-Time policy, quality assurance or co-development of new products. However, if a Supply Chain Management approach creates long-term partnerships because it is a source of higher quality, the buyer must not forget his role as a prospector in his constant search for suppliers that are best suited to accompany the company's strategic evolution;

– net requirements calculation through the Material Requirement Planning method, in direct connection with production. The net requirements calculation allows you to refine the supply quantity and the resulting inventory management (see Sheets 19–22 and 32);

– management of orders to suppliers through purchase requisitions, calls for tenders, specifications, contracts, orders, receipts, disputes and invoices. Although considered operational, the first steps up to the drafting of the contract are essential, because it is these documents that will regulate all future agreements (and disagreements);

– the prospective management of a stock (see Sheets 19–22).

Purchasing and Logistics also take part in the ecological approach, in the implementation of *eco-sourcing* ([CHA 11], see Sheet 15).

"Two points of attention" of purchases are proposed below, both of which the logistician must be aware, namely "the *Bullwhip* effect" and the call for tenders, followed by the specifications and the contract (see Sheet 24).

4.1.2. *The Bullwhip, "whiplash" or Forrester effect*

Sometimes, in the relationship between operators in the supply chain, the choice is to call on operators (suppliers and sometimes producers) who are distant from the customer or end consumer. Involving remote operators can be the result of a financial decision (lower costs) or the absence of a specialist in the desired product in the area close to the sale (or production). In this context, however, risks are multiplied in terms of information loss and/or increased damage. This is called the Bullwhip effect. This effect is related to the magnification of the variations in the relationships between operators, as we move away from the customer or end consumer. Thus, the greater the risk of relying on distant suppliers and/or producers, the greater the risk grows. Similarly, the further away we get from the consumer, the more stock will have to be held in order to limit (1) stock-outs and (2) transport costs and customs duties. Finally, this pattern leads to a variation in the production program, which may have an impact on smoothing policies in production (see Sheet 32).

It was in 1958 that Jay Forrester demonstrated the phenomenon of *"amplifying the variability of demand"*, which was later referred to as the Bullwhip or Forrester effect. *"It starts with simple modeling [...] by relying on the systemic principle of feedback and using several variables: stock, personnel, production rate, delays and orders. This modeling then makes it possible to decide, given these initial conditions, how many people should be hired the following week. However, due to delays and the need to project into the future, production does not adjust seamlessly... and it can be observed that a slight change in demand leads to an oscillatory behavior of the response given. In short, even with constant demands, repeated local adjustments can lead to instability of hires and multiply, in a purely endogenous way, the effect they are supposed to correct".*

Box 4.1. *The Bullwhip or Forrester effect (source: extract from [ROU 16])*

4.1.2.1. *Example of the Bullwhip effect on orders and production launch*

The Bullwhip effect generates significant variations in order management and, in turn, in the launch of production.

Imagine a manufacturer whose production site is located in Italy and who distributes its products to central warehouses in the capitals of Germany

(Bonn), Austria (Vienna) and France (Paris). The central warehouses then distribute the regional stores weekly with stable quantities. Deliveries from central warehouses to regional stores are as follows:

– 300 boxes per week from Bonn;

– 600 boxes per week from Vienna;

– 900 boxes per week from Paris.

Note that the shipment is made up of pallets of 60 boxes and that a complete truck carries 30 pallets.

The Italian producer delivers to the warehouses with free shipping (that is, the buyer does not pay transport costs) if the order reaches a full truckload of 30 pallets.

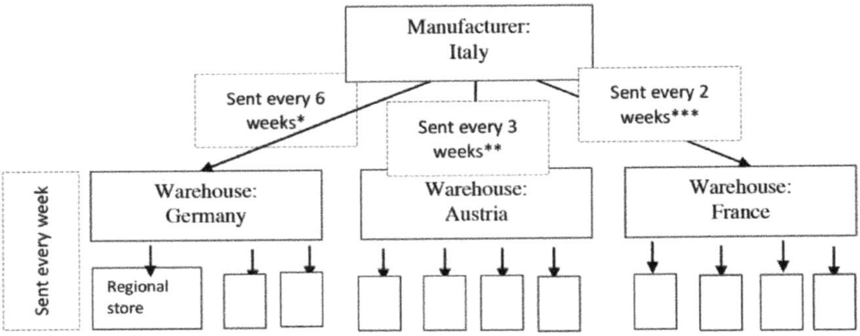

Legend:
**300 boxes / 60 boxes = 5 pallets per week ➔ 1 truck = 30 pallets ➔ 30 / 5 = one shipment every 6 weeks from the Italian producer to the central warehouse in Germany.*
***600 boxes / 60 boxes = 10 pallets per week ➔ 1 truck = 30 pallets ➔ 30 / 10 = one shipment every 3 weeks from the Italian producer to the central warehouse in Austria.*
****900 boxes / 60 boxes = 15 pallets per week ➔ 1 truck = 30 pallets ➔ 30 / 15 = one shipment every 2 weeks from the Italian producer to the central warehouse in France.*

Figure 4.1. *Graphical representation of the Bullwhip effect*

The following additional data is provided to establish the delivery schedule and launch of production. The demand d for pallets per week is 5 for Germany, 10 for Austria and 15 for France. The stock at the beginning of the period is 0 for each warehouse. The production launch takes place 1 week before the scheduled delivery. The table is drawn up over a period of

2 months. To facilitate the understanding of the exercise, we will consider <u>deliveries to be made at the beginning of the week.</u>

	Week	**–1**	**0**	**1**	**2**	**3**	**4**	**5**	**6**	**7**	**8**
Germany St. 0 = 0 d = 5	St.	0	25	20	15	10	5	0	25	20	10
	Liv.		30						30		
Austria St. 0 = 0 d = 10	St.	0	20	10	0	20	10	0	20	10	0
	Liv.		30			30			30		
France St. 0 = 0 d = 15	St.	0	15	0	15	0	15	0	15	0	15
	Liv.		30		30		30		30		30
Total outflows from the Warehouses to the Stores*			30	30	30	30	30	30	30	30	30
Total outflows from the Producer to the Warehouses**			90	0	30	30	30	0	90	0	30
Production launch***		90	0	30	30	30	0	90	0	30	?

Legend:

St. = Stock; d = demand for pallets per week; Liv. = Delivery

**each week, the stores in Germany are delivered with 5 pallets, 10 pallets for Austria and 15 pallets for France, that is, a total of 30 pallets per week.*

***this corresponds to deliveries by a full truck from the producer to central warehouses in each country.*

****the launch of production at the producer's premises takes place 1 week before.*

Table 4.1. *Delivery and production launch table – the Bullwhip effect*

Explanation of Table 4.1:

– delivery from the Italian manufacturer takes place at the beginning of the week in the central warehouse (arrow down);

– delivery from the central warehouse to the regional stores takes place during the week (arrow up);

– the total outflow from warehouses to stores is 30 pallets per week, that is, 5 for Germany, 10 for Austria and 15 for France;

– the total number of deliveries from the Italian producer to the central warehouses varies per week, depending on shipments from the central warehouses to the regional stores.

The example shows that, depending on store demand, the variability of demand varies to a greater or lesser extent: one shipment every 6 weeks for Germany, one shipment every 3 weeks for Austria and one shipment every week for France. It can also be seen that the impact for the manufacturer – both in terms of consignments and production – is subject to considerable variation and affects the search for a smoothing of production.

4.2. Sheet 24: Tendering Process, Specifications and Logistics Service Contract

Key point

Considering the variety of logistics services, Carbone [CAR 04, p. 115] noted that it is difficult to "draw up a legal archetype". In fact, it is the terms of specifications and timetable for implementation that must be clearly identified. Hence, the importance of the logistician's role in this drafting of specifications, considering sometimes the high technical nature of the services requested.

4.2.1. *Tendering process and specifications*

The specifications are a document that describes what the project manager must do in relation to the client's requirements. It serves as a contractual basis between a contractor and a bidder in the framework of tendering. The call for tendering may be private or public. It involves hiring a number of candidates to meet the company's needs. It is called "open" when any operator can answer it and "restricted" when only candidates selected by the buyer can answer it. In the context of public procurement, Order no. 2015-899 of July 23, 2015 specifies the procedure for invitations to tender.

The drafting of specifications prior to the contract is therefore essential. According to the AFNOR standard [AFN 91] NF X 50–151, the functional specification is *"the document by which an applicant expresses his needs (or those he is responsible for translating) in terms of service functions and production constraints. Each function has its own assessment criteria and levels, with a certain degree of flexibility attached to each"*.

4.2.1.1. *Specification examples*

The French Standard EN 16271 of February 2013 specifies *"the expression of needs and functional specifications"*. As for us, we present below a list of points to consider when drawing up specifications. This list is therefore not exhaustive and does not take the specific constraints of certain trades and products into account (see Table 4.2).

Points	Description
The problem	This first point describes the problem to be addressed. It presents the context, locating where this problem needs to be solved. It highlights as precisely as possible the malfunction(s) with which the company is confronted.
Constraints	Constraints can be physical, geographical, social (e.g. adaptation to users' skills) and in terms of access security (e.g. military zone and computer access).
Functionalities	If the product already exists, the existing one should be described in addition to the functionalities expected.
The deadline for completion	A GANTT diagram (see Sheet 39) that specifies the date of completion of each step is recommended.
Who does what?	What is the responsibility of the client and the contractor? If there is already knowledge of trade experts, this should be specified. There is also the possibility here of indirect operators in fulfilling the need, but from whom it is necessary to transmit information on site monitoring, for example.
Confidentiality	It must be dealt with in the broadest sense, including intra- and inter-company IT.
Conditions of receipt	At this stage, all the implementations are to be mentioned, in terms of pre-testing (or pre-receipt), guarantees, unpacking and installation and security. We also need to think about future updates. If these are not foreseen from the outset, a new contract will have to be drawn up each time.
Specific clauses	This point can specify everything related to after-sales service and employee training in the present and also in the future; in addition, the obligation to hold a certain standard, etc.

Table 4.2. *Specification examples*

Des Mesnards [DES 14] proposed a structure of functional specifications that can also be used in a logistics framework. Table 4.3 presents an adaptation.

Title	Sub-title	Description
1. Presentation of the project for the design, re-design and improvement of a product or service		
1.1 The context	1.1.1 The facts	– The current situation, how unsatisfactory it is, how it motivates the project;
	1.1.2 The causes	– the reasons for current dissatisfaction;
	1.1.3 The aims	– the issues sought through the project.
1.2 The objective and expected results		– Specify that it is a design, re-design or improvement, and what product or service it is. Expected results are indicators of project success.
1.3 Stakeholders		– The applicant, the decision maker (who makes the final development choices, which may be different from the applicant); other stakeholders involved in the project.
1.4 The perimeter		– The scope of the project, the elements to be treated and not to be treated.
1.5 The constraints to be respected		– Time frame, budget, obligations (technical, commercial, financial, regulatory, etc.).
1.6 Criteria for choosing solutions		– Criteria according to which the choice will be made between several solutions meeting the objective and respecting the constraints. The satisfaction of needs and cost is essential criteria.
2. Identification of the product or service concerned		
2.1 The product or service		– Designation.
2.2 Main user and main characteristics		
2.3 Environment of the product or service		– Life cycle; – interactions with the environment (human, tangible, intangible, etc.); – conditions of use.
3. Service functions		
3.1 Types of functions		– Function of use; – function of estimation (example: aestheticism); – adaptation function (example: shock resistance).
4. Realization constraints		
4.1 Constraints		– General constraints; – specific product or service constraints.

Table 4.3. *Content of specifications (source: adapted from [DES 14])*

4.2.2. *The logistics service contract*

After the specification comes the contract. In drawing up a logistics service contract, Carbone [CAR 04] stressed that it cannot be equated with an employment contract (there is an independence of the service provider from the client) or a contract of deposit (there is performance of a task). This type of company contract therefore follows the provisions of the general legal regime of the Civil Code, Article 1787 et seq[1].

Item	Designation
1st	Purpose and scope of application.
2nd	Definitions on the following points: the parcel, the freight forwarder, the principal, dispatch, delivery, goods, ancillary services, pick-up and reservations.
3rd	The obligations of the principal in relation to the information and documents to be provided by the principal to the freight forwarder, verification of documents, illicit or prohibited goods, transport equipment and reserves.
4th	Packaging and labeling of goods – reporting obligations.
5th	The obligations of the freight forwarder with regard to his substitutes, in terms of authorizations for enforcement, criminal liability, transmission of information and documents, control, etc.
6th	Ancillary services.
7th	Goods insurance.
8th	The delivery.
9th	Failure of the principal, impediment to transport.
10th	The price of services.
11th	The modification of the current contract.
12th	Terms of payment.
13th	Responsibility.
14th	Prescription.
15th	The duration and termination of the commission contract.
16th	The jurisdiction clause.

Table 4.4. *Items in a standard contract of a transport commission (source: extract from Decree no. 2013-293 of April 5, 2013, repealed on May 28, 2014)*

There are two types of contractual relations between logistics and industrial service providers, namely the "closed book" relationship (the service provider's remuneration is based on the volume processed) and the

1 https://www.legifrance.gouv.fr/affichSarde.do;jsessionid=821EA65E3BE09E74789D6B8A89450EC4.tpdila15v_1?reprise=true&page=1&idSarde=SARDOBJT000007118002&ordre=null&nature=null&g=ls

"open book" relationship (the service provider's remuneration is based on a previously defined fee) [ROQ 03].

There is no standard logistic service contract. However, we propose the example of the standard transport commission contract below, as proposed by Decree no. 2013-293 of April 5, 2013 – a version revoked on May 28, 2014[2].

4.3. Sheet 25: The InCoTerms

Key point

The InCoTerms for *INternational COmmercial TERMS*. In France, Incoterms are also called C.I.V's or Conditions Internationales de Vente *(International Sales Conditions)*. InCoTerms represent the obligations of the buyer and seller under an international contract. InCoTerms are determined at the time of the contract signature, that is, at the first stage of the supply chain. InCoTerms determine the allocation of costs between buyer and seller, but do not determine the cost of transporting goods as such.

4.3.1. *Functions of the InCoTerms*

Three essential functions are attached to InCoTerms:

– determine the place of delivery of the goods at an agreed place. This step is associated with the implementation of administrative documents whose purpose is to enumerate all the elements and stages that condition the shipment;

– define the respective obligations of the seller and the buyer for the shipment of the goods. Here, we specify who is responsible for loading and unloading, which type of transport is used, the insurance to be taken into account and the distribution of transport costs;

– allocate the costs and risks of the services performed for each delivery. In other words, if a damage or other type of problem occurs when the goods are transferred, the place where the damage or problem occurred specifies the operator who will bear the damage.

---

2 https://www.legifrance.gouv.fr/affichTexte.do?cidTexte=JORFTEXT000027276884&dateTexte=20140527

There are four categories of InCoTerms corresponding to a decreasing level of obligations for the seller and an increasing one for the buyer. These categories are represented by the first letter from InCoTerms: E, F, C and D. For example, InCoTerm EX-Work gives the seller the least amount of obligation. He positions only his goods on the shipping dock and the buyer takes care of all aspects of the shipment (transport, customs, insurance) until arrival.

4.3.2. *The different InCoTerms*

The current list of 11 InCoTerms is listed on the Government Customs website[3]. This list of 2010 InCoTerms was applied on January 1, 2011.

4.3.2.1. *InCoTerms for transport by sea and inland waterways*

Four InCoTerms can be used for transport by sea and inland waterways (see Table 4.5).

InCoTerm	Description
FAS or Free Alongside Ship	*The seller pays the transport costs to the shipment port, carries out the export formalities and pays the costs, duties and taxes related to these formalities. The goods are delivered along the ship in the shipment port: this delivery marks the transfer of risks and costs. The buyer shall bear the costs of loading, maritime transport, unloading and transport from the port of destination to its place of activity.*
FOB or Free On Board	*The goods are delivered to the ship designated by the buyer. Under the 2010 rules, the notion of a rail crossing that had previously materialized the transfer of risk has disappeared. From now on, the transfer of risk and costs takes place when the goods have been delivered to the ship. The seller pays the transport costs to the shipment port, as well as the loading costs, carries out the export formalities, and pays the costs, duties and taxes related to these formalities.*
CFR or Cost and Freight	*Risk is transferred when the goods are delivered on board the ship to the shipment port. In addition, the seller shall bear the transport costs to the port of destination and the loading costs, he shall carry out the export formalities and pay the related duties and taxes. The buyer shall bear the costs from the moment the goods arrive at the port of destination, carry out the import formalities and pay the duties and related taxes.*
CIF or Cost, Insurance and Freight	*Risk is transferred when the goods are delivered on board the ship to the shipment port. In addition, the seller shall bear the costs of transport and insurance to the port of destination and the costs of loading, he shall carry out the export formalities and pay the related duties and taxes. The buyer shall bear the costs from the moment the goods arrive at the port of destination, carry out the import formalities and pay the duties and related taxes.*

Table 4.5. *InCoTerms for transport by sea and inland waterways (source: http://www.douane.gouv.fr)*

3 http://www.douane.gouv.fr/articles/a10836-incoterms-pour-une-meilleure-performance

4.3.2.2. *InCoTerms for all types of transport*

Seven InCoTerms can be used for all types of transport (see Table 4.6)

InCoTerm	Description
EXW or Ex-Works	*The goods are available at the seller's premises on a fixed date. The buyer organizes and pays for the transport. He also bears the risks up to the final destination of the goods. The buyer is also responsible for the export and import formalities and costs, as well as the duties and taxes related to these two operations.*
FCA or Free Carrier	*The seller shall deliver the goods to the carrier designated and paid for by the buyer. The transfer of risk is materialized during this operation. Export formalities and costs, as well as related duties and taxes, are to be carried out by the seller. The buyer assumes the transport to these places of activity, carries out the import formalities and pays the related duties and taxes.*
CPT or Carriage Paid To	*The seller shall bear the costs of shipping to the port of destination. The transfer of risk is established when the goods are made available to the first carrier. Thus, the buyer is responsible for the insurance costs.*
CIP or Carriage and Insurance Paid To	*The conditions are the same as for CPT. The seller must provide insurance cover for the buyer, the risk of loss and damage that the goods may incur during transport.*
DDP or Delivered Duty Paid	*The goods are delivered at the place of destination, ready to be unloaded, while the seller has cleared the goods for export and import, and paid duties and taxes related to these operations. In principle, the buyer shall bear the costs of unloading, unless the contract stipulates that the seller is responsible for such unloading. At the same time, the buyer is under no obligation to pay for inspection costs, while the seller pays for pre-shipment inspection costs, which are required by the authorities of the exporting or importing country.*
DAT or Delivered At Terminal	*In addition to the main transport, the seller arranges and pays for the unloading of the goods at the destination and their transport to the agreed terminal. The transfer of risk is effective when the goods have been made available to the buyer at the agreed terminal. The buyer carries out the import formalities and pays the duties and taxes due as a result of the importation.*
DAP or Delivered At Place	*The seller is responsible for the transport of the goods to the agreed delivery point, so he bears the costs and risks up to this point. The goods are made available to the buyer at destination on the means of transport, without being unloaded. The buyer organizes the unloading, carries out the import formalities and pays the duties and taxes due as a result of the importation.*

Table 4.6. *InCoTerms for all types of transport (source: http://www.douane.gouv.fr)*

4.3.3. *The Baltic Dry Index*

InCoTerms are not intended to determine the cost of transport. However, in the context of the cost of shipping dry bulk goods (minerals, coal, cereals), the Baltic Dry Index (BDI), published by the Baltic Exchange in London and in circulation since November 1 1999, makes it possible to specify charter rates.

4.3.4. *Packing list*

In addition, within the context of shipments, the logistician must transmit a packing slip, which is an *"international trade document that is always in the documentary bundle and on which all packages constituting a shipment are listed. This list makes it possible to verify the conformity of the shipment with the order, in other words, the invoice established by the seller. It indicates the identification of each parcel (brands, numbers, weight, etc.) and also gives the total of the shipment (number of parcels, cubage, gross weight, etc.) and the packaging used. It is a document required by customs, as well as on importation"*[4].

4 https://www.glossaire-international.com/pages/tous-les-termes/liste-de-colisage.html

5

Production Logistics

5.1. Sheet 26: Key Concepts of Production Logistics

Key point

Production logistics, sometimes called internal logistics, is central in companies, especially in the secondary or industrial sector. The most extensive work on maximizing production rates is formulated at this level. The challenge in production logistics is to match production rates with those of distribution, while remaining flexible and responsive. This requires a detailed calculation of production requirements and planning.

5.1.1. *Criteria in production*

Three criteria in production can be used (see Table 5.1).

Criteria	Type	Description
Technique	Continuous process	Manufacture of products without any interruption (e.g. oil and chemicals).
	Discontinuous process	Manufacture of products individually, by means of a chain (e.g. textile).
	Process by project	Manufacture of a single product for a limited time (e.g. boat)
Quantity produced	Unit production	The choice of the quantity produced is generally based on technical and/or cost criteria. The more technical and/or expensive the product is, the more unitary the production will be.
	Batch production	
	Large series production	
Relationship with the customer	Stocked production	This concerns the so-called push-flow production.
	Firm Order Production	This concerns the so-called pull-flow production.

Table 5.1. *Criteria in production*

5.1.2. *Workflows in production: push, pull and tight flows*

In production, first, we will talk about push-flow production. In this context, production requirements drive the entire supply chain. Physical flows and information flows go in the same direction, from upstream to downstream. We talk about the production of standard goods, which are all the same. Production usually takes place in large quantities, in bulk. The presence of stock is omnipresent, and in order to regulate these stocks, the policy is to implement as many deliveries as possible in fixed quantities and on a regular basis.

Then, the production can be pull-flow. The production process is triggered by the customer's request, and is often associated with the payment of a deposit on an order. The production order is therefore established in relation to the customer's requests: the information flows from downstream to upstream. We are talking about the so-called personalized assets.

However, few logistics chains can have a 100% pull-flow policy, perhaps for luxury goods and more! Still, businesses are often competing and have to find ways to be the first choice for consumers. Among the methods, it has become a prerequisite to play around with deadlines by shortening delivery times. In addition, companies are confronted with market needs where customization is often required, but for which the desire to obtain is almost immediate. The solution then goes through a balance between "push flow" and "pull flow": in the push flow, the maximum number of parts of the product is produced and assembled, in its common composition and via the pull flow, the customer's order – through customization – is finalized. The point of equilibrium between push flow and pull flow is called the Customer Order Decoupling Point or delayed differentiation (see Figure 2.1).

There is also a third type of flow in production: the tight-flow method. This method aims to minimize inventory and adapt production to demand. In its optimum state, we are talking about a pull-flow production. This production method is often associated with the "just-in-time" logic.

5.1.3. *Just-in-time (JIT)*

"In 1912 in Meiji, China, Taiichi Ohno[1] *was born, the man who probably exasperated the world's production engineers the most! All of them owe him the 'just-in-time', this method specially developed to reduce production costs. It is based on a rational organization of assembly lines, known as Kanban, which involves reducing stock, production by type of object, labeling the product to track its history, etc. A graduate of Nagoya University in Japan in 1932, Taiichi Ohno developed his method at Toyota, where he made his first steps in the textile industry. In 1943, he joined the automotive subsidiary as inventory manager. The president of this company, Kiichiro Toyoda, declared in August 1945: 'It is absolutely necessary to catch up with the United States, otherwise the Japanese automobile industry will die'.*

"He exasperated engineers and workers.

"At the time, Taiichi Ohno notes that the production of an American worker was nine times that of a Japanese worker. Inspired by the work of Professor Shigeo Shingō and after a study trip to the United States, Ohno devised his 'just-in-time' method in the 1950s. It was adopted by Toyota in 1962, but was not immediately applied: the engineer had to face the reluctance of workers, who had to adapt. The efficiency of the 'just-in-time' approach would, however, appeal to the entire world industry from the 1973 oil crisis onwards. As vice-president of Toyota from 1975 to 1978, Taiichi Ohno remained as a group advisor until 1982. He died in 1990 in Toyota City, Japan."

Box 5.1. *Just-in-time (source: extract from Finyear Daily newspaper:* finance & innovation all the year[2], *accessed 05.29.2017)*

According to Lamouri and Thomas [LAM 99], "*Just-in-time leads to the elimination of all waste. This means that 'if you don't need something now, then don't do it'*". According to these two authors, three foundations support the JIT, notably: "*proper stock management, good cost control and a progress policy*". For this, it is necessary to produce in small batches, implementing the principles of total quality and preventive maintenance, and lastly, to work on firm orders.

Lamouri and Thomas [LAM 99] enumerated the tools to be associated with just-in-time, while stressing that if one of these tools is not used

1 Also spelled Taiichi Ōno.

2 https://www.finyear.com/Le-juste-a-temps-1960-Just-in-time_a247.html.

because the company's resources are not suitable, this does not mean that we should not apply a just-in-time policy:

– Kanban (Japanese term meaning *"sheet" or "label"). Lyonnet and Senkel [LYO 15] related that "the Kanban method is a production planning tool that helps to limit waste due, in particular, to overproduction and unnecessary handling. [...] it is an information system materializing the order of a customer workstation that is located upstream of the workflow. Thus, the quantity produced by the upstream workstation will be limited to the actual requirements of the downstream workstation"*;

– the SMED or Single Minute Exchange of Die translates as "Matrix change(s) in one minute" and its purpose is to "*minimize the downtime of a production means to switch from one production to another*";

– make fluid production through the establishment of workshops (see Sheet 29);

– implement Total Quality Management (see Sheet 39).

> Shigeo Shingō is a Japanese engineer who designed the Kanban production method, the SMED system and the Poka-Yoke method.

Box 5.2. *Shigeo Shingō (1909–1990): designer of Kanban, SMED and the Poka-Yoke Method (source: Wikipedia extract)*

5.2. Sheet 27: Lean Manufacturing Fundamentals

> **Key point**
>
> Lean manufacturing is an approach that was developed by Toyota and made public by the *Massachusetts Institute of Technology* (MIT) in the 1980s. The principles of the *Toyota Production System* (TPS[3]) have taken the common name of lean manufacturing. Recognized, this method consists of producing as efficiently as possible, by eliminating all waste with the aim of achieving "*industrial excellence*" [PET 06]. Petitqueux [PET 06] pointed out that lean manufacturing is conceived through a number of foundations and tools. In this sheet, we present the six foundations, then the tools in the following sheet.

3 Taiichi Ohno (see Box 5.1) and Eiji Toyoda (1913–2013) instigated the TPS.

5.2.1. *Industrial excellence*

Industrial excellence is based on the stability of resources such as products, costs, processes, trust, management, etc.

> *On May 30, 2017, the 23rd edition of the 2017 factory trophies was held on the seven categories of industrial excellence: (1) productivity (Pouyet 3M company), (2) quality (Bic company), (3) processing (Delacre Nieppe company), (4) CSR*[4] *initiative (Adam company), (5) the Human Resources initiative (Schmidt company), (6) customer experience (L'Oréal company), and (7) the factory (Bosch Mondeville company).*

Box 5.3. *Industrial Excellence Awards 2017 (source: https://www.entreprisedufutur.com/evenements/trophee-des-usines-7-trophees-pour-recompenser-l-excellence-industrielle)*

According to a study by Chick *et al.* [CHI 14], companies that have won an industrial excellence trophy have implemented one or more of the following strategies:

1) *"They use data and information flows to facilitate the close integration of their partners in the supply chain;*

2) *They optimize the value offered to the customer throughout the entire chain, and not just in the segment that concerns them*[5]*;*

3) *They cooperate with suppliers to rapidly improve their manufacturing processes;*

4) *They use their technical knowledge to offer their customers a highly personalized product".*

Industrial excellence is achieved through the partnership of supply chain operators, enabling them to implement innovative solutions and thus stand

4 CSR stands for corporate social responsibility; it was originally created in the 1950s to limit some unethical excesses of the organizational economic system, and it specializes in the social aspect by advocating the voluntary integration of social and ecological concerns in industrial and sales activities [JOR 02].

5 For example, this point can be applied by the implementation of the Efficient Consumer Response approach (see Sheet 33) or even a logistics of returns (see Sheet 35).

up to the so-called low-cost companies: "*the result: together, they have developed innovative solutions capable of offering the end user a better product and service. This is an approach that emerging countries' low-cost factories cannot easily replicate, as it requires sharing proprietary data, optimizing the overall costs of the value chain, rapidly adapting a factory's production capacity by integrating suppliers and being able to manufacture highly customized products. All these are trump cards that European industrialists have at their disposal in order to succeed in the future*" [CHI 14].

5.2.2. *The notion of waste*

The notion of waste is seen through the elimination of the 3Ms, which are *Muri*: irrational situations, *Mura*: variations and *Muda*: tasks with no added value or waste.

Unfortunately, Lorino [LOR 14] argued that the use of the 3Ms has evolved, removing the quintessence of this approach and no longer doing it justice. Thus, for this author, there is a "before" and an "after" in lean manufacturing. The "before" is part of "*an organizational model that continuously redesigns the collective system of activity (hence the central reference to the Muri – to workload, that is, effort) to improve its ability to create value*", whereas the "after" is mainly devoted to eliminating Muda (waste) at the expense of the other two: "*Twenty years later, it is striking to me how hard we have worked to eliminate wastage (Muda) and how little we cared about the Mura (irregularity) and Muri (overload). In short, irregularity and overload are now the root causes of waste in many organizations. 'Worse still, they bring waste back to where managers and operational teams thought they had eliminated it' [WOM 06]. The 2013 'lean new look' thus turned out to be a fairly certain path to failure*" [LOR 14].

5.2.3. *Kaizen through continuous improvement*

Kaizen means the search for continuous improvement or "analysis to make better". It is the fusion of the Japanese words Kai for "change" and Zen for "good" or "better". Masaaki Imai (1930–present) instigated the Kaizen in the 1980s and founded the Kaizen Institute Consulting Group (KICG) in 1986.

Régol and Bélanger [RÉG 03] presented the seven principles of Kaizen, which are (1) the cycle of research, design, production and sales that underline the importance of good interaction between services; (2) Total Quality Control at all levels such as products, services, personnel, costs, security, etc.; (3) a wide dissemination of reliable data; (4) client satisfaction in terms of quality, cost and deadlines; (5) managing problems upstream through teamwork; (6) prevention through suggestions for improvement; and (7) quality circles.

As for Suarez-Barraza *et al.* [SUA 11], the Kaizen is part of three perspectives: a philosophical management, a component of Total Quality Management and a theoretical principle for improving methodologies and techniques. Many principles and tools are proposed by these authors, including:

– Gemba's management improvement. Gemba, or Genba, is a Japanese term meaning "where reality lies". In lean manufacturing, this term means "field', "workshop" or "workstation". Instead, we are talking about "doing Gemba Walks" or Genchi Genbutsu, which means "*going to the source to verify the information for yourself, to make sure you have the right information to make the right decision. [...] Gemba presents current problems that we have to face, while Genchi Gembutsu is more focused on seeking opportunities for improvement that we need to find*" [LOP 15];

– total productive maintenance. Created in Japan in 1971, Total Productive Maintenance is a proactive approach to improving the efficiency of machines through an evolution of maintenance methods. Seiichi Nakajima (1919–2015) is considered to be the pioneer of this system;

– Jidoka consists of "*reducing the number of operators needed to monitor the machines. And this is the strength of automation, which could be best understood by the 'automatic stop'; when a machine starts to notice that the parts it produces are of poorer quality, that it overheats or that it lacks oil, it stops automatically by activating an andon, which will alert the maintenance worker. This means that you only need one operator where you needed 4 or 5 before. These systems that are self-automated are called Poka-Yoke (or 'coded', or 'error proof' is a device, usually mechanical, which avoids assembly, installation or connection errors); they confer machine intelligence*" (source: Wikipedia);

– the SMED for Single Minute Exchange of Die translates into a change of press tool per minute (see also Box 24.2);

– the PDCA and SCDA cycle: the PDCA cycle for *(1) Plan, (2) Do, (3) Check, (4) Act* or (1) Study/plan/analyze a work, (2) Execute/implement, (3) Review, check the result in accordance with what was intended, (4) Adjust if the result is not consistent or standardize if the result is consistent. This is also called the Deming Wheel (see Sheet 39). SDCA stands for *Standardize, Do, Check, Act*. The SDCA is placed before the PDCA cycle. Its purpose is to take stock of the existing situation. Two other approaches are similar to the PDCA and SCDA cycles: PRTS for *Plan, Try, Review, Standardize* and DMAIC for *Define, Measure, Analyze, Improve, Control*;

– DMAIC for *Define, Measure, Analyze, Improve, Control* and DOE for *Design of Experiments*.

5.2.4. *The Heijunka or the smoothing of production (see Sheet 32)*

Inspired by the Toyota Production System, the Heijunka consists of smoothing the production in order to make maximum use of equipment during off-peak periods, in addition to avoiding overtime during peak periods. This approach also considers the product mix, that is, seeking a balance of production according to the specifications of the products. This means that production is not necessarily started by order of arrival, but by grouping together products from the same family.

5.2.5. *The just-in-time*

Just-in-time (JIT) corresponds to the *"ideal rate at which the production system must operate in order to meet demand within a minimal time frame"* [PET 06]. It is presented and developed in Sheet 26.

5.2.6. *The Jidoka*

The Jidoka has the objective of developing machines and processes capable of detecting any anomaly. It is also called self-automation and it is cited by Suarez-Barraza *et al.* [SUA 11] in section 5.2.3.

5.3. Sheet 28: Lean Manufacturing Tools

Key point

Sheet 27 presents the six lean manufacturing fundamentals as cited by Petitqueux [PET 06]. In addition to these, this author lists lean manufacturing tools such as value stream mapping, the operations diagram, the 5S approach, delayed differentiation, the SMED method, assembly line, workstation ergonomics, unit production workflow, Poka-Yoke, Kanban, the "small train" and "short interval animation".

5.3.1. *Value stream mapping*

Value stream mapping (VSM) is considered to be the "best" mapping of a production chain, both quantitatively and qualitatively [SER 08][6]. The construction of a VSM uses standard pictograms for easy reading. VSM particularly breaks down the tasks of the production process. It highlights two main elements, namely the process steps that are usually represented by machines and the "running time" between and during machine operation. For each machine, a technical data sheet provides key indicators for evaluating the performance of the machine.

The preparation/establishment of technical data sheets for each stage of the industrial process is a key point of a VSM. Several indicators are used. Table 5.2 presents the main indicators and their interpretation.

---

6 In addition to the VSM, Serrano Lasa *et al.* [LAS 08] listed the following as tools for mapping the productive system: (1) process mapping, (2) the Icam Definition Zero (IDEFO) method, (3) the Graphs with Interrelated Results and Activities (GRAI) method, and (4) the modeling and simulation software for material and information flows.

Indicators	Description
Breaks Time	Breaking time
BS – Batch Size	Batch size: typical batch size that is started in process as a unit
C/T – Process Cycle Time	Process cycle time: time to complete a production part (by unit)
C/O – Change Over Times & Frequency	Change time and frequency
Demand	Average number of units per change
EPE – Every Product Every	Production lot size
FTT – First Time Through	From the very first time: percentage of jobs that are complete and precise from the first time they were launched into production
MR – Machine Reliability	Machine reliability
No. Ref	Order number
No. Shift	Number of changes
OEE – Overall Equipment Effectiveness	OEE – overall equipment effectiveness: machine utilization rate
Process ratio	Process ratio: total process time/throughput time
Scrap rate	Scrap rate
Take time	Rate of demand: number of pieces requested by the customer, per unit of time: (net transaction time/period)/(customer requests/period)
Throughput time	Sum of delay times and process time
Uptime	Usage: percentage of machine time available for production
Value-Added Ratio	Value-added ratio: total value added/operating time
Workers	Workers (in numbers[7])
Workers times (minus break)	Working time (less stoppages)

Table 5.2. *Indicators used in VSM*

5.3.2. *The operations diagram*

The operations diagram is a graphical representation of the different operations in a process. It thus makes it possible to highlight non-value-added shares that can therefore be canceled and conversely for those that require clarification. The operations diagram can be assimilated to a logigram or flow diagram (see Sheet 39).

7 It is permissible to list employees as "full time equivalent" or FTE, in other words the number of employees is reduced in relation to the legal working time, that is, 35 hours per week.

5.3.3. *The 5S approach*

The 5S's step-by-step approach refers to five Japanese terms: *Seiri* to get rid of the unnecessary, *Seiton* for tidying, *Seiso* for cleaning, *Seiketsu* for maintaining order/standardizing and *Shitsuke* for being rigorous/maintenance and improvement.

5.3.4. *Delayed differentiation*

It is on the writings of Tarondeau [TAR 98] that a majority of authors rely for the delayed differentiation which "*consists of delaying the point of differentiation of the product or process (from which each product acquires its own identity within a range), in order to supply semi-finished products rather than finished products. The goal is to produce as many standard elements as possible and to push back the point where each product is different from the others and needs to be identified as such*" [AGA 04].

Several other terms are related to delayed differentiation, such as the order penetration point or Customer Order Decoupling Point, post payment, order penetration point [GIA 07] or co-manufacturing: "*Co-manufacturing corresponds to a delayed differentiation approach whereby the PSL [logistics service provider] will receive, on behalf of an industrial customer, basic industrial components for which it ensures final assembly or assembly at the latest, depending on firm orders from customers*" [FUL 05].

5.3.5. *The SMED method*

The SMED method (for Single Minute Exchange of Die) translates as "matrix change(s) in one minute" and aims to "*minimize the downtime of a production means in order to switch from one manufacturing process to another*". It was designed by Shigeo Shingō in the 1970s (see Box 5.2).

5.3.6. *The assembly line*

When Petitqueux [PET 06] presented the "assembly line", he referred to the search for a reduction in waste, such as waiting times, transport, trips, etc.

In 2007, Lasnier remarked: "*Some recommendations call the design habits of production lines into question. It is now a question of working without a large safety stock between workstations, but with stocks of components at the workstation, eliminating the need for operators to move around, implementing U-shaped lines, and being versatile and autonomous on several workstations. Flexibility is required so as to eliminate waiting times, and the chain of operations and movements are carried out in takt time, that is, at a given rate*".

5.3.7. *Ergonomics of the workstation*

Ergonomics of the workstation consists of adapting the machines, information, environment, etc. from work to the employee. Several guides and trainings are proposed and recommended to help in this approach of work. For example, the *ministère de la décentralisation et de la fonction publique* (Ministry of Decentralization and Public Service) offers a 65-page practical guide on the prevention of musculoskeletal disorders in a 2015 edition[8]. Reflection on the ergonomics of a workstation must involve teams of employees concerned, occupational physicians, ergonomists and members of the CHSCT[9].

5.3.8. *The unitary production flow*

According to Petitqueux [PET 06], the unitary production flow consists of setting up a production where each upstream operation continuously feeds the next operation. Therefore, any production that has been started must be finalized. This occurs either when the product is completely standardized or if the order is personalized when the firm order reaches the company.

8 Available at: https://www.fonction-publique.gouv.fr/files/files/publications/coll_outils_de_la_GRH/guide_pratique_TMS.pdf.

9 The CHSCT or *Comité d'Hygiène, de Sécurité et des Conditions de Travail* (Health, Safety and Working Conditions Committee) is compulsory for all establishments with more than 50 employees (source:http://travail-emploi.gouv.fr/sante-au-travail/acteurs/comite-d-hygiene-de-securite-et-des-conditions-de-travail/qu-est-ce-qu-un-chsct/article/le-comite-d-hygiene-de-securite-et-des-conditions-de-travail-chsct).

5.3.9. *The Poka-Yoke*

The term Poka-Yoke or Poka-Yoké is a Japanese term which translates as "error-proof system" or "coded". This method was introduced in 1961–1963 by Shigeo Shingō in a philosophy of respect for people and their intelligence: "*In a race for repetitive operations on which vigilance and memory depend, the poka-yoke can save time and free the human spirit to more creative operations and sources of value creation*" [DUD 09]. The purpose is therefore to set up control and warning systems, which can go as far as to stop production if necessary.

5.3.10. *The Kanban*

Initiated in the 1950s by Shigeo Shingō and applied by Taiichi Ohno in the just-in-time approach, Kanban – which translates as "label" or "sheet" – is a sheet that is placed downstream of the machine and "goes up" when the need for a production launch is felt. Lyonnet and Senkel [LYO 15] specified the six rules of its use:

– *"The forward item goes upstream to remove the exact quantity on the Kanban;*

– *the backflush manufactures the withdrawn quantity as it appears on the Kanban sheet and in the order of withdrawals;*

– *nothing must be produced or transferred without a Kanban;*

– *a Kanban must accompany the products;*

– *production must be stopped immediately when a faulty part appears;*

– *absolute perfection must be required at every workstation, and the number of Kanban in circulation must be as small as possible".*

5.3.11. *The "small train" or Mizusumashi*

Under different names, the "small train", Mizusumashi, milk run or water spider consists of placing the parts requested by the operators on a moving vehicle. This vehicle passes along the machines and allows operators to pick up the necessary parts without having to move around.

5.3.12. *Short interval animation*

By "short interval animation", Petitqueux [PET 06] means the implementation of frequent reportings, the purpose of which is to be aware of the intended and achieved level of performance. For example, you can find control charts and other quality tools (see Sheet 39), as well as dashboards (see Sheet 41).

5.4. Sheet 29: Implementation of Workshops

Key point

The establishment of production workshops is necessary for a better fluidity of flows. However, setting up a workshop is a complex practice, because it forces the logistician to take many criteria into account, such as the size of the building and machines, the type of machinery, the layout of the traffic lanes, possible storage areas and safety zones. This is a complex practice. In addition to the search for a better fluidity, it is necessary to think about reducing delays when switching between the different machines.

5.4.1. *Implementation methods*

In his doctoral research, Hachicha [HAC 09] proposed a state of the art on *cell formation methods* that presents the implantation methods used for the organization of a workshop. There are two types of methods: those based on the shape of the products and those based on the production range of the different products:

– the classification and coding method is mainly part of the product form-based methods. Codes are associated with the product, highlighting the nature of the raw material, shapes, tolerances, etc. The machines are grouped together, taking into account the maximization of common product characteristics;

– several sub-methods are included in the methods based on the production range of different products. Table 5.3 summarizes these methods.

Methods	Description
Array-based Clustering	*"consists of [...] arranging the order of rows and columns to obtain a matrix of blocks along the diagonal. These blocks identify the groups of machines forming cells and the families of product that can be produced by these cells"* [HAC 09]. The best-known methods are those of Kusiak [KUS 87] and King [KIN 80].
Hierarchical grouping	*"is based on the definition of a coefficient of similarity or distance between products and machines. The objective is to classify similar elements (products and machines) together according to their similarities"* [HAC 09].
Mathematical programming	mainly consists of heuristics: *"the number of cells and the number of machines and products affected for each cell are determined by matrix manipulations"* [HAC 09].
Graph theory	in which particularly with the Bipartite graph method, "one node type represents products and the other type presents machines" [HAC 09]. The objective is to reduce the number of machine arcs/products and network flow approaches.
Non-hierarchical grouping	contains the ideal seed non-hierarchical clustering (INSC). Here, we are looking for "*a performance indicator called grouping efficiency that measures intercellular movement and machine utilization*" [HAC 09]. In addition to the INSC method, the author cites the ZODIAC model and the GRAFICS method as improvements to the original INSC method.
Automatic classification	is the creation of aggregates to the "nearest neighbor". The algorithm of dynamic clouds is classified there.
Artificial intelligence	includes calculations based on optimization algorithms, evolutionary approaches, neural networks and fuzzy logic.
Heuristics	consists of reducing as many possibilities as possible.

Table 5.3. *Methods based on the range of different products produced (source: adapted from Hachicha [HAC 09])*

5.4.2. *Main types of implementation*

In general, four types of workshop implementation are considered, namely:

– random implementation, usually in small productions;

– the "job shop" layout, in other words functional or homogeneous sections. Machines of each family are put together;

– the "flow shop" layout. The machines are installed according to the product flow rate, in other words they are installed according to the manufacturing process of the product. The methods recommended in this type of implantation are the average rank method and the anteriority method;

– the so-called "island" implantation. This concerns the specialized manufacturing process: each product batch has its own set of machines. Hachicha [HAC 09] presented it under the title of Array-based Clustering (see Table 5.1). Three methods are associated with this type of implantation: the King method [KIN 80], the Kusiak method [KUS 87] and the chain link method. Madani [MAD 14] considered that King's method is more rigorous than Kusiak's; however, it requires a specific table or software such as IMPACT[10].

Whatever the type of installation chosen, a precise analysis must be carried out upstream on the environment (size, workshop layout) and on the specification of the machines. Concerning the specification of the machines:

– if, at any time in the production process, a cooking step is used, that is, in the use of an oven, it can only be implemented under certain safety guarantees. This specification can sometimes modify the implementation logic, as in the case of a "flow shop" implementation;

– by the same logic, if the process uses a cooking machine, it must be remembered that when using this type of machine, this step cannot be stopped at all. In fact, only if the product allows it should a buffer zone for storing the product upstream of this machine be considered when this cooking stage occurs at the closing of the workshop. For example, the product requires two hours of cooking and the workshop closes at 5 p.m, but the product arrives at the cooking stage at 4 p.m, in which case the "cooking" stage cannot be started. In short, the logistician must associate the "establishment of a workshop, specification of the product and scheduling constraints"; but the product must also be able to wait overnight to begin the cooking stage;

– the size of the machine is also a criterion to be taken into account before installation. Thus, from the very beginning, it is necessary to think about replacing every machine because of breakage or antiquity. This means that it is preferable – if possible – to place very large and very heavy machines close to the access point, so that they can be replaced without having to completely move all of the other machines.

10 The IMPACT software was designed and is distributed by CLIO Annecy Transfert (source: http://qlio-annecy-transfert.com/logiciel/impact/).

5.5. Sheet 30: Production Planning: the Business Plan

Key point

Production planning takes place over a time horizon ranging from daily (day to day) to long term (more than two years). Production planning manages machines, materials and people according to sales forecasts, with the aim of maximizing the use of these three types of resources on a daily basis. In production planning, the logistician also seeks to "smooth out production".

5.5.1. *Levels of production planning*

Starting from a long-term planning horizon and moving on to daily management of the scheduling flows of operations, five production planning levels are cited (see Table 5.4).

Horizon	Level	Description
two years or more	Business plan	Carried out per strategic area of activity
Up to one year but can be revised during the year, e.g. every three months	S&OP (sales and operations plan)	Is generally carried out by technological family In direct connection with the sales plan (sales department) and stock management decisions
Monthly or even weekly	MPS (master production schedule)	Carried out per finished product Includes S&OP business data
Monthly or even weekly	MRP (material requirement plan)	Is carried out on the materials and components of each product
Every day	Scheduling operations	Includes, but is not limited to, work planning, scheduling, start and follow-up of production

Table 5.4. *The different levels of production planning*

5.5.2. *The business plan*

Every business strategy is part of a planning process. Johnson *et al.* [JOH 05] noted that "*strategic planning aims to develop and coordinate an organization's strategy through systematic, orderly and sequential procedures*". According to Stutely [STU 12], "*a business plan presents the current situation of the organization and defines the overall strategy of the company for the next five years, based on a detailed operational and financial plan for the coming year. The strategy and plan encompass all aspects of the business. The most important questions vary from one company to another, but in general we focus on management, products, marketing and sales*".

A business plan is developed by a strategic business unit (SBU). A company can have 1 to n SBU, which is a subpart of a company dedicated to a specific product/market segment. This means that each SBU has its own dedicated technological, human and financial resources.

As a whole and according to Stutely [STU 12], a business plan must combine a strategic plan, a marketing plan, an operational plan, a financial plan and possibly a construction project. To guarantee its success, Stutely considered that 10 + 2 steps are necessary, namely (1) to define the company's activities, (2) to define the current situation of the company, (3) to define the external market, the competitive positioning and the company's market, (4) to define the objectives for the whole period covered by the business plan, (5) to develop a strategy to achieve the defined objectives, (6) to identify risks and opportunities, (7) to develop a strategy to limit risks and exploit opportunities, (8) to operationalize strategies, (9) to forecast expenses and revenues, and create a financial plan, (10) to draft the financial plan in a concise manner, (11) to get the business plan approved and finally (12) to use it!

The implementation of a business plan requires transversal skills where logistics operations are omnipresent. Consequently, the logistician must understand a number of strategic tools for consolidating a business plan. We will only quote the main tools here.

Strategic tools can be grouped into two categories: those analyzing a sector of activity from a global point of view (macro-environment) and those focusing on the enterprise itself (micro-environment).

5.5.2.1. *The main macro-environmental strategic tools*

The main strategic tools for assessing the behavior of companies in a sector of activity are the following:

– PESTLE for Political-Economic-Social-Technological-Legal-Environmental. The PESTLE analysis consists of listing the influences that affect a particular industry sector. For example, in the pharmaceutical sector, the Legal category of PESTLE lists all the laws, decrees, etc. that affect this sector. It is possible to refine a PESTLE by specifying the "weight" of each item (how does this item affect the work of each company in this sector of activity: recommended – mandatory?) or the "impact" over time (if we know that a law will be introduced in the coming months or years, it is useful to specify this temporality as soon as a PESTLE is drawn up);

– PORTER's five forces model[11]. This model assesses the bargaining power (1) of customers and (2) of suppliers; the threat (3) of potential market entrants and (4) of alternative products and services; and (5) the competitive intensity of the sector. Without Porter's agreement, a sixth force was added to this model, namely the influence of the public authorities. Finally, in the graphical representation of this model and on a scale from 0 to 10, we speak of the sectoral hexagon.

5.5.2.2. *The main micro-environmental strategic tools*

The two main strategic tools for assessing a company's situation are:

– the SWOT model for Strengths, Weaknesses, Opportunities and Threats. Its conception is based on the LCAG structure named using the creators' initials: Learned, Christensen, Andrews and Guth [LEA 65]. At the time of analysis, this model lists the company's main strengths and weaknesses as well as the market's threats and opportunities, and the company's potential for action. The list of threats and opportunities can be based on the PESTLE tool. As for the strengths and weaknesses, and in order to not forget some of them, it is recommended that these two categories be

11 Michael Porter (1947–present) is an American professor of business strategy. He is famous for designing a number of strategic tools used to diagnose the behavior of companies within a sector of activity. Among these tools, the Five Forces model and the generic strategies are the most cited. There are three generic strategies [POR 86]: "cost dominance" (covering a large audience with low product sales prices, mainly consumer goods), "differentiation" (the audience is large, but selling prices are higher, including electrical and electronic equipment) and "focus" or "niche strategy" (the audience concerned is small and selling prices are more or less varied; there are luxury products).

undersized into four aspects, namely the commercial aspect (analysis of the product, distribution, communication and price), the financial aspect (debt, cash flow, liquidity, etc.), the technological aspect (innovation potential, ownership of patents, production costs, productivity, etc.) and the organizational aspect (type of management, control system, workforce, etc.);

– the Porter value chain [POR 86]. Porter's value chain is the identification and analysis of a company's activities. In its original model, Porter considers two types of activities, namely the main activities (procurement logistics, manufacturing, commercialization, marketing and sales, after-sales service) and support activities for the main activities (infrastructure such as finance, human resources, research and development, and purchasing).

5.6. Sheet 31: Production Planning: Sales and Operations Planning

Key point

S&OP, short for sales and operations planning, is the second stage of production planning (see Table 5.4). The S&OP takes into account the sales forecasts of the sales department and the quantities that can be produced. S&OP is based on families of products and volumes, not references. It is consolidated by the general management, in view of the data provided by the sales and financial departments.

5.6.1. *The relationship between sales forecasts and production potential*

The first step in developing the S&OP is to determine sales forecasts issued by the sales department. These sales forecasts will result in the necessary production. With consideration of this data, the production department will formulate a potential production according to its production constraints (machine and human productivity). The objective is to try to bring the sales forecasts as close as possible to production, while taking three important constraints into account, namely maximum production, storage capacity and the management of heavily detailed salary costs (standard hours vs. overtime, temporary staff, etc.).

5.6.1.1. *Development of sales forecasts*

The sales department considers three criteria for a sales forecast:

– first, an analysis of sales by product using a "time series". The time series consists of looking at the sales behavior over several years, quarter by quarter – or month by month if it is a product with strong seasonality. This analysis assesses the sales trend of the product and whether it is growing, stagnating or declining. However, in order to ensure that a time series can be interpreted, it is recommended that past sales should be adjusted for any exceptional and non-renewable events, such as a climatic event that boosted or canceled sales, or the shutdown of machines that modified production;

– second, customer segments. There are four types of customer segments, namely (1) the current customers of the company whom it is trying to keep, (2) the current customers of the competition whom it is trying to recover, (3) relative non-consumers, who are being targeted to become current customers (advertising, etc.) and (4) absolute non-consumers who, no matter what is done, will never buy the product sold by the company;

– finally, the general management strategy, which may decide to expand and therefore increase sales, or conversely shut down one or more production lines.

The analysis and extrapolation of these three criteria allow for the forecasting of future sales and production requirements.

5.6.1.2. *Development of the quantities to be produced*

Based on sales forecasts, the Production Department prepares the production quantities. Here too, however, considerations must be taken into account in order to check the feasibility of this production. Thus, a production required for sales forecasts may sometimes not be compatible with the production requested and will therefore require adjustments.

The elements valued by the logistician in determining a production are:

– the establishment of a safety stock. The product for sale can be added as a safety stock. This quantity is therefore added to the production program;

– raw material requirements. The logistician must consider the scrap rate. The presence of scrap rates therefore means that 100% of a raw material may not be used. This means that more raw materials must be purchased to account for this waste. The issue of using raw materials particularly arises in stages 3 and 4 of industrial planning, namely the Production Plan and the Net Requirement Calculation;

– the actual rate and time of use of the equipment. Machinery adjustment, material loading, sampling, incidents, cleaning, maintenance, breaks, stops, etc. must be deducted from the operating time of an equipment. From these cutbacks, the logistician must also assess the "waste rates" relating to technical losses (falls, breakage, etc.) and quality defects. This is referred to as the "machine utilization rate" or sometimes "machine productivity";

– the actual working time of the operators. Paid leave, public holidays, absences due to accidents or illness, preparation time (dressing, etc.) and breaks (meals, etc.) must be taken into account. The term used for this is "staff productivity".

The combination of criteria makes it possible to determine the possible production. This production is compared with the production demanded through the sales forecast, and it is this comparison that sheds light on whether the company has the ability to respond to sales forecasts or whether, on the contrary, it is necessary to use overtime and/or recruit more personnel or invest in more equipment.

5.6.2. *Example of a sales and operations plan*

A sales and operations plan is a document that consists of a forecast sales plan, a production plan and a stock plan.

The forecast sales plan is given for the year 2018 for a product group. A safety stock is decided on in order to deal with forecast errors.

	Jan.	Feb.	Mar.	April	May	June	July	Aug.	Sept.	Oct.	Nov.	Dec.
Sales forecast	900	720	455	1260	1782	1782	1311	900	2064	1765	1650	1800
Safety stock	180	141	90	249	354	354	261	180	411	351	330	360

Table 5.5. *Fiscal year: sales forecasts and safety stock*

	Number of days worked	Number of operators	Possible Prod.*	Cumulative availability **	Sales forecast	Cumulative forecast	Remaining stock***	Safety stock	RS – SS (4*)
IS				450					
Jan.	20	20	1200	1650	900	900	750	180	570
Feb.	20	20	1200	2850	720	1620	1230	141	1089
Mar.	23	15	1035	3885	455	2075	1810	90	1720
April	21	15	945	4830	1260	3335	1495	249	1246
May	18	15	810	5640	1782	5117	523	354	169
June	22	25	1650	7290	1782	6899	391	354	37
July	19	25	1425	8715	1311	8210	505	261	244
Aug.	10	25	750	9465	900	9110	355	180	175
Sept.	23	30	2070	11535	2064	11174	361	411	**–50**
Oct.	21	30	1890	13425	1765	12939	486	351	135
Nov.	19	30	1710	15135	1650	14589	546	330	216
Dec.	22	30	1980	17115	1800	16389	726	360	366

Legend: IS = initial stock

**The production of an operator is three units per day. Monthly production is obtained by multiplying by three the number of days worked and multiplying that by the number of operators.*

***The cumulative availability takes into account the total production of the previous month plus the monthly production. Note that for January, the initial stock is taken into account.*

****The remaining stock is calculated as the difference between the cumulative availability and the cumulative (sales) forecast. We see here that the storage constraint of 2000 units is respected. No remaining stock exceeds 2000 units.*

*****Stock remaining – safety stock. The security stock constraint for September is not respected. There are 50 pieces missing. This means that 50 pieces will have to be produced in addition, which will modify the S&OP at the end of the year.*

Table 5.6. *Fiscal year: the sales and operations plan*

Production constraints are established such that:

– the production rate is three units/day/operator;

– the number of operators dedicated to the production of this family of products is consolidated such that: January and February: 20 operators; March, April and May: 15 operators; June, July and August: 25 operators; and from September to December: 30 operators;

– the opening stock at the beginning of January is 450 units;

– storage constraints are 2000 units.

The number of days worked is provided on the result table and for ease of reference, the data provided in the practice already take the wastage rates and wastage times for machinery and men into account, and no differentiation between normal hours and overtime is considered.

The S&OP is thus established:

	Number of days worked	Number of operators	Possible Prod.*	Cumulative Availability**	Sales forecast	Cumulative forecast	Remaining stock***	Safety stock	RS – SS (4*)
IS				450					
Jan.	20	20	1200	1650	900	900	750	180	570
Feb.	20	20	1200	2850	720	1620	1230	141	1089
Mar.	23	15	1035	3885	455	2075	1810	90	1720
April	21	15	945	4830	1260	3335	1495	249	1246
May	18	15	810	5640	1782	5117	523	354	169
June	22	25	1650	7290	1782	6899	391	354	37
July	19	25	1425	8715	1311	8210	505	261	244
Aug.	10	25	750	9465	900	9110	355	180	175
Sept.	23	30	2120	11585	2064	11174	411	411	0
Oct.	21	30	1890	13475	1765	12939	536	351	185
Nov.	19	30	1710	15185	1650	14589	596	330	266
Dec.	22	30	1980	17165	1800	16389	776	360	416

Table 5.7. *Fiscal year: the sales and operations plan, adjusted for the constraint of the safety stock over the month of September*

5.7. Sheet 32: The Production Plan: the Master Production Schedule – Material Requirement Plan – Production Smoothing

Key point

The MPS (master production schedule), the mrp (material requirement plan) and production smoothing are the third and fourth steps in production planning (see Table 5.7). These planning stages are based on a monthly or even weekly horizon.

5.7.1. *The relationship between the sales and operations plan and the master production schedule*

The sales and operations plan (see Sheet 31) is not based on references, but on product families. Conversely, the master production schedule involves references. However, these two plans combine, in the sense that the MPS translates in terms of reference what the S&OP has consolidated in terms of product family.

The MPS therefore determines production requirements by reference. To do this, the logistics specialist determines the raw requirements per finished product and sub-sets, taking the lot size[12], supply and production constraints, stock and the start dates of each production step into account.

5.7.2. *The architecture of a master production schedule*

To calculate net requirements, an architecture for the production of a finished product on the different workshops is recommended. This architecture makes it possible to visualize both the production process between workshops and the nomenclatures, that is, the quantities required for each material and the launch times between workshops.

For example, take the finished Product A. In order to produce one unit of A, one unit of F and one unit of G are required. In the intermediate workshop that manages the sub-sets, it is noted that one unit of F is composed of one unit of X, one unit of Y and one unit of Z, and that one unit of G is composed of one unit of Y and two units of Z.

It can also be noted that the launch time between the intermediate workshop and the final workshop is one month and the launch time between

12 It is possible to be confronted with allotment constraints, that is, the release of a batch of products is done by quantity and not by unit, which can change the production quantities. For example, a machine can be programmed to run only batches of 10 units. However, allotment constraints are usually found at the last stage of the production process, in other words when it is placed on the pallet. Thus, a pallet can "impose" on a load quantity.

machining and the intermediate workshop is one month for F and two months for G. Finally, sub-set Z is purchased directly, machined from an external subcontractor. The lead time between the intermediate workshop and the machine shop is one month for all X and Y products, and also one month for the externally acquired Z product.

The architecture of the master production schedule is established as shown in Figure 5.1:

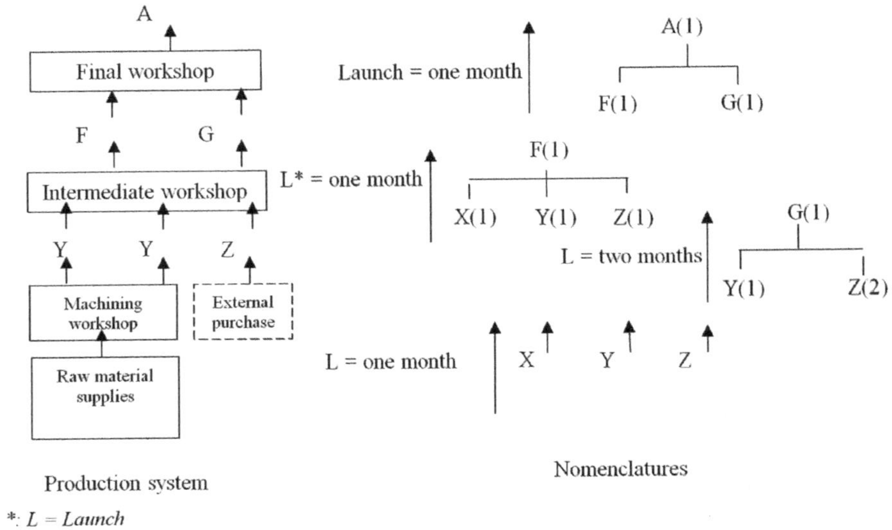

Figure 5.1. *Architecture of the production system and nomenclatures*

5.7.3. *Determination of net requirements*

The MRP (material requirement planning) calculation takes into account the capacities of the machines (and possibly lot-sizing constraints), the presence of safety stock and the release time between production and delivery. It adapts the production – if necessary – according to sales forecasts, seeking a "smoothing of the production".

The following six months' worth of data are provided:

	January	February	March	April	May	June
MPS forecast* Finished Product A	10900	10800	10100	12500	12400	10800
MPS – Spare part F	1500	1600	2100	1800	1800	1500
MPS – Spare part G	1700	1600	1800	1500	1400	1700
Inventory position end of month Dec.	**A**	**F**	**G**	**X**	**Y**	**Z**
	400	500	700	500	1000	300
Expected deliveries (current stock available)	**A**	**F**	**G**	**X**	**Y**	**Z**
January	10500	27400	15000	35000	34200	41000
February			12000			

**MPS: master production schedule.*

Table 5.8. *Master production schedule and stock data*[13]

Launch calculations in terms of the final workshop and intermediate workshop are established such that:

Finished Product A	December	January	February	March	April	May	June
Gross requirements (MPS)		10900	10800	10100	12500	12400	10800
Expected delivery		10500					
Stock at end of month	400	0	0	0	0	0	0
Net requirement			10800	10100	12500	12400	10800
Launch (one month)		10800	10100	12500	12400	10800	?

The calculation of the final stock is as follows:

Final stock (FS) = initial stock*+ expected delivery - gross requirements

For January, we have FS = 400 + 10500 - 10900 = 0

For February, FS = 0+0-10800 = - 10800 => the negative figure then corresponds to net requirements and, of course, there is no stock left

The launch takes place with a delay of one month

**The initial stock (IS) of the current month corresponds to the final stock (FS) of the previous month.*

Table 5.9. *Finished Product A launch chart*

13 Note that in the example provided, no safety stock is considered. If this is the case, the launch calculations will have to be adjusted if the final stock obtained does not cover the safety stock.

Sub-set F	December	January	February	March	April	May	June
Gross requirements from A*		10800	10100	12500	12400	10800	?
Spare part F		1500	1600	2100	1800	1800	1500
TOTAL		12300	11700	14600	14200	12600	?
Expected delivery		27400					
Stock at end of month	500	15600	3900	0	0	0	
Net requirements		0	0	10700	14200	12600	?
Launch (one month)		0	10700	14200	12600	?	

For January, FS = 500 + 27400 - 12300 = 15600

For February, FS = 15600+0-11700 = 3900

For March, FS = 3900+0-14600 = -10700 => the negative figure then corresponds to net requirements and, of course, there is no stock left

**The gross requirements for assembly F arise from the use of this assembly in finished product production. Here, the finished Product A requires one sub-set of F (see section 5.7.2). Therefore, F's gross requirement corresponds to the number found on the launch line of the finished Product A[14].*

Table 5.10. *Sub-set F launch chart*

Sub-set G	December	January	February	March	April	May	June
Gross requirements from A		10800	10100	12500	12400	10800	?
Spare part G		1700	1600	1800	1500	1400	1700
TOTAL		12500	11700	14300	13900	12200	?
Expected delivery		15000	12000				
Stock at end of month	700	3200	3500	0	0	0	
Net requirements		0	0	10800	13900	12200	?
Launch (two months)		10800	13900	12200	?	?	

Table 5.11. *Sub-set G launch chart*

The same type of calculation/reasoning will apply for raw materials X, Y and Z.

14 If, for example, the nomenclature had specified that two units of F were required for a finished Product A, the launching quantity of A would have to be multiplied by this number two to have the gross requirements of F, and so on.

5.7.4. *Production smoothing*

Production smoothing is necessary in two cases. First, when the production capacity of a workshop is limited and then when the operators are vacant over a given period of time.

In the proposed example, the production capacity of the intermediate workshop is limited. Monthly capacity in number of hours is 127 for January, 182.5 for February, 184 for March and 122 for April. We also know that the production capacity in one hour is 200 units for F and 100 units for G. In view of this information, it is possible to obtain the operating times and to answer any problems of adjustment concerning "load – capacity".

Intermediate workshop	January	February	March	April
Launch in F Launch in G	0 10800	10700 13900	14200 12200	12600 ?
Charging in hours for F* Charging in hours for G**	0 108	53.5 139	71 122	63 ?
TOTAL LOAD	108	192.5	193	?
TOTAL CAPACITY (see statement)	127	182.5	184	122
CAPACITY – LOAD	+ 19	- 10	- 9	
Final launch of G	10800 + (19 h * 100 p.) = 12700	13900 – (10 h * 100 p.) = 14900	12200 – (9h * 100 p.) = 13100	

**Charging in hours for F (see Table 5.10) = 0/200 for January; 10700/200 for February, etc.*
***Charging in hours for G (see Table 5.11) = 10800/100 for January; 13900/100 for February, etc.*

Table 5.12. *Load and capacity table for production smoothing*

Table 5.12 shows a non-use of machines for 19 hours in January, even though there are not enough hours for February and March. In this case, a "smoothing operation" can be planned. When a production smoothing operation is carried out, we go from the future (April) to the past (January). In the example, we see that 10 hours are missing in February and 9 hours in March, while there are 19 unused hours in January[15]. It is possible to transfer

15 In the example, there is compensation between surpluses and production shortfalls, which is rarely the case in reality. Where production compensation or smoothing is not possible, overtime will be required in order to meet the requested production. As a reminder, a production must always be paralleled with the storage capacities, and even more so in "production smoothing" operations.

the missing hours from February and March to January and smooth the production.

To determine the product on which the load and capacity adjustment is made, the reference with the lowest possible hourly stock value is determined. The hourly cost of an outstanding amount includes material costs, variable charges (for example, electricity) and fixed charges (for example, electricity subscription, fixed salaries)[16]. In this case, the hourly cost of the lowest outstanding amount is considered to be for the sub-set G. The adjustment is therefore made on it.

16 The difference between a variable load and a fixed load is that whatever the production, a fixed load is always paid, whereas the variable load depends on the production carried out. Thus, all that is covered by subscriptions, rents, depreciation and amortization or the fixed portion of salaries are expensed as fixed charges.

6

Downstream or Distribution Logistics

6.1. Sheet 33: The Problem of Downstream or Distribution Logistics

Key point

Downstream or distribution logistics is the logistics that manages the relationship with the end consumer. It is the (last) link in the supply chain, which was very early on considered to be crucial. Indeed, it is the one that must deliver the right product to the end customer very quickly. It is therefore directly linked to market constraints and demands. The word "last" is put in parenthesis here, because the evolution of logistics is such that the delivery of a finished product is no longer considered as the "end" of a supply chain. In fact, reverse logistics (see Sheet 35) can give rise to other logistics chains, such as waste reprocessing.

6.1.1. *The problem of downstream or distribution logistics*

Downstream logistics is the logistics close to the customer. In this respect, it is the first to pass on information in order to specify the type of production process to be implemented. Thus, distribution logistics is directly affected by market constraints and realities. It makes it possible to specify whether the demand (or the sales forecast) is based on logistics in push flow or pull flow. It therefore specifies the life cycle of orders.

Downstream logistics works closely with customers. To respond quickly to their requests, there are several procedures and techniques to manage the flow of information and goods. These approaches and techniques include ECR for efficient consumer service, SSM for shared supply management and CPFR for collaborative planning, forecasting and replenishment.

When talking about the consumer and/or the end customer, several operators are involved. There is the individual buyer of single or small quantities of parts, and also there are wholesalers (see Sheet 9) and retailers. According to the operator, we are talking about Business-to-Consumer – BtoC (in the first case) – and Business-to-Business – BtoB (in the other cases).

Lastly, it can be said that downstream logistics ensures the availability of the product to the customer, based on an analysis of the process of (re-) supply that requires:

– definition of the structure of the circuit and distribution channel;

– definition of the inventory management and replenishment systems for finished products;

– definition of the working procedures in the warehouses (see Sheets 17 to 22) and the organization of delivery rounds;

– processing orders, handling, packing, storage, warehousing, stock management, transport, shipping, if necessary.

6.1.2. *The efficient consumer response (ECR) approach*

The efficient consumer response (ECR) approach was initiated in the 1990s in the United States. The implementation of this partnership approach starts from the food distributor to the industrial supplier and seeks to save costs through a better relationship [VAN 98]. ECR's objective is to establish cooperation between supply chain operators [BON 99, PAC 97, DES 00, BON 05].

http://ecr-france.org/qui-sommes-nous

ECR, an active network of exchanges and reflection promoting collaboration, the sharing of experiences and good practices, brings together decision-makers from manufacturers, distributors and logistics service providers in the food and non-food consumer goods sector.

ECR is the acronym for efficient consumer response, which stands for the joint industry-to-trade initiative, aimed at providing an optimal response to the consumer:

Supply chain: (1) anticipate supply chain transformation; (2) define appropriate supply processes to maximize customer service rate; (3) improve organizational efficiency;

Consumer offer: (1) improve the monitoring of sales by panels for each distribution channel; (2) optimize the effectiveness of promotions; (3) understand and anticipate new consumer-shopper practices.

http://ecr-france.org/trophees-ecr

Winners 2016

Gold Trophy: AUCHAN RETAIL and L'OREAL for their dossier "a partnership at the service of linear availability with the C2C project "Collaboration to Consumer"".

Silver Trophy: METRO and COCA COLA Company for their dossier "increasing recycling rates in the out-of-home segment".

Bronze Trophy: SAVENCIA the Proxi Visiomodular concept of Savencia for the cheese department.

Startup ECR Trophy - Jury prize: ELEMENTUM

Startup ECR Trophy - ESCP Student Award: ELEMENTUM

Box 6.1. *ECR France (source: http://ecr-france.org/)*

6.1.3. *SSM, CPFR and MMS*

Several collaborative practices have emerged along the downstream supply chain and are part of ECR's approach. We quote:

– Shared supply management (SSM): *"Shared supply management (SSM) is a form of collaboration between distributors and their industrial suppliers that is currently developing strongly. SSM is based on intensive electronic data interchange (EDI). The distributor, or the logistics service provider to whom warehouse management has been entrusted, shall communicate data daily to the supplier, describing the withdrawals for the categories of products concerned. This information enables the supplier to establish forecasts and prepare an order proposal, which is sent to the distributor's central purchasing department for validation. The flows are thus driven by demand. This technique is well suited for mass-market consumer products that are delivered frequently (one or more times a week)"* [FEN 05]. For GS1 France, SSM is *"a logistics collaboration process between an industrialist and a distributor, where the replenishment activity is shared between the two partners and driven by demand. From the warehouse stock movements (SSM warehouse) or cash withdrawals (SSM point of sale) transmitted by the distributor, the industrialist takes charge of his customer's supplies by calculating and sending an order proposal"*. SSM initially leads to a relationship between 1 distributor and 1 supplier.

– Vendor managed inventory (VMI) or supplier managed inventory is a variant of SSM. In this context, the vendor completely manages his stock on behalf of his customer. For example, the vendor alone decides on delivery quantities and shipping schedules.

– Co-managed inventory (CMI) or collaborative managed inventory is also a variant of SSM and VMI. It is also called JMI for Jointly Managed Inventory. Here, however, the vendor merely proposes a delivery schedule and delivery quantities, and the vendor starts the replenishment procedure with the customer's prior approval.

– The multipack SSM consists of using several vendors to replenish distributors. This approach allows better loading in terms of transport means and, as a result, a reduction in cost through this massing of deliveries.

– Multi-drop SSM is the "opposite" of multipack SSM. A single vendor delivers to several distributors.

– Collaborative planning, forecasting and replenishment (CPFR®) is related to SSM but the CPFR also includes promotional operations. The CPFR is a registered trademark of the Voluntary Interindustry Commerce Standards (VICS) from 1988[1].

1 https://www.gs1us.org/DesktopModules/Bring2mind/DMX/Download.aspx?Command=Core_Download&EntryId=492&language=en-US&PortalId=0&TabId=134

– MMS or mutualized management of supplies is defined as "a mode of supply management in which several manufacturers undertake delivery to one or more distributors together from the same logistics site (warehouse or platform), in order to optimize storage costs (reduction of the number of storage sites in the industrial-distributor chain) and transport costs (filling rate of vehicles and number of rotations)" [LIV 06]. MMS is the equivalent of multipack SSM. It is also referred to as mutualized SSM or pooling (see Box 6.2).

6.1.4. *The case of e-commerce*

Talking about distribution logistics also requires an interest in e-commerce. E-commerce or electronic commerce is part of the Business-to-Consumer relationship. Through e-commerce, consumers define and place their orders via their computers.

For the logistician, two configurations of order picking can (co-)exist, namely (1) dedicated on-site preparation, which is located in a warehouse where the activity is mainly focused on digital orders and (2) in-store preparation, whose electronic activity is often more marginal but which grows with the drive[2] implementation [PAC 08, DUR 10].

According to Pierre Fournet (associate director of Diagma), MMS can be implemented in various forms with the aim of increasing delivery frequencies, reducing inventory levels and improving quality of service:

– MMS in the form of mutualized SSM (MSSM): here, several manufacturers practicing SSM (shared supply management) with the same distributor group their goods in the same full truck. Specialized software (EWR or OCS[3]*) managed by a*

2 "The term "drive" generally refers to a collection point for goods or merchandise, where the customer takes delivery of the goods directly into or near the vehicle. The term "drive" was originally used to refer to the possibility of ordering and retrieving from drive-by's available in fast-food chains. It has also been used for a few years now to designate the points of withdrawal offered by supermarket chains that are predominantly food retailers, and allow the delivery of items to the trunk of the vehicle. In this case, the order is placed beforehand on the Internet from a computer or mobile application on a smartphone and the customer chooses a withdrawal time slot" (source: https://www.definitions-marketing.com/definition/drive/).

3 EWR and OCS are two pieces of software designed by Generix and CTS respectively: *"At the origin of shared computing, there is the OCS (Optimization of Orders and Inventories) software package, designed and built by the CTS (Centre Télématique Services) to meet the requirements of Mars Unisabi in search of SSM"* (source: http://www.faq-logistique.com/Communique-presse20100605-ASLOG-Carnet-Thematique-SI-OCS-CTS.htm).

3PL or 4PL[4] *ensures the management and optimization of flows. The mutualized SSM allows an excellent optimization of truck filling, because it is possible to adjust the quantities of each supplier's SSM orders.*

– MMS in the form of a collaborative SSM. The industrial sector practice SSM with their distributor customers. The objective is to satisfy all the SSM orders calculated. This makes it possible to use the same truck for several suppliers. Truck filling optimization is much lower than in MSSM because the quantities of SSM orders cannot be adjusted. Jean-Luc Fonteneau believes that "we should be able to pool without practicing SSM". *Hence the other modes of pooling.*

– MMS in the form of opportunity-based SSM: this is the same principle as collaborative SSM, except that manufacturers are not in SSM mode. The objective is to group orders from several suppliers in the same truck.

– MMS in the form of mixed SSM, combining manufacturers supplying their customers with SSM and other manufacturers receiving a classic order from the distributor customer. In concrete terms, we start by loading the classic orders, and complete the truck with pallets in SSM.

Box 6.2. *The different forms of MMS (source: extract from http://www.aslog.org/fr/404-diagma-fait-la-promotion-de-la-gma.html?input2=&input1=&mots=&nb_res=0&niv2=25)*

DOSSIER - E-commerce continues to grow in France. Sales in the sector reached 72 billion euros in 2016 and are expected to reach 80 billion euros in 2017. The expansion of supply and the development of marketplaces continue to drive consumption.

In 2016, purchases made by French people on the Internet reached 72 billion, after 65 billion in 2015 and 57 billion in 2014, according to Fevad, the Fédération de l'e-commerce et de la vente à distance *(Federation of e-commerce and distance selling). A figure up by 11% year-on-year. As in previous years, e-commerce continued to be driven by the expansion of supply – with the number of merchant sites increasing by 12% in 2016 to 200,000 – and by the development of marketplaces, which continued to gain ground. The only drawback is that the average basket amount, at 70 euros, has fallen by a further 7% over one year, but the average number of transactions per buyer per year has increased. Moreover, e-commerce still generated only 8% of retail sales in 2016, which leaves room for*

4 Logistics service providers: 3PL – third-party logistics; 4PL – fourth-party logistics (see Sheet 10).

growth. In 2017, Fevad is targeting a total of 80 billion euros in sales. The 100 billion mark should be crossed by 2020.

Box 6.3. *The rise in power of e-commerce (source: https://www.lesechos.fr/industrie-services/dossiers/0201064436807/0201064436807-la-montee-en-puissance-du-commerce-en-ligne-165287.php – 26/01/2017)*

6.2. Sheet 34: The Warehouse/Platform Alternative in Distribution Circuit

Key point

Logistics must be aware of the distribution circuits it faces. In fact, the typology of the circuits influences the modes of delivery and storage. The alternative for the company is then to choose between a warehouse (with storage) and a distribution center (without storage).

6.2.1. *The difference between channel, circuit and distribution network*

According to the classical approach:

– The distribution channel corresponds to the number of intermediaries present in the distribution of products. We are talking about three types of channels: the direct channel where sales are made directly, without intermediaries (manufacturer => end consumer), the short channel, where an intermediary appears (manufacturer => wholesaler or retailer => end consumer), and the long channel, which includes several parties (manufacturer => wholesaler => retailer => end consumer).

– The distribution circuit or formula takes all the distribution channels used by a certain category of products into account.

– The distribution network gathers the stakeholders who will deliver the product to the end consumer (e.g. manufacturer, wholesaler, retailer).

6.2.2. *Selection criteria for the location of a repository*

When the distribution circuit is determined, the logistician has to choose the type of depot best suited to respond to product deliveries. However, before building a repository, the following questions must be answered:

– How many warehouses/platforms are needed?

– Where should they be located?

– What should the size of the depot(s) be?

The answers to these three questions are subject to the following issues:

– Overall costs of transporting the goods to the customer. Routing costs take all transport costs to the consumer into account. However, they may also consider the costs of emptying or reprocessing waste[5].

– Load breaks. Load breaks correspond to the loading and unloading of goods in terms of the mode of transport. The logistician seeks to reduce breakdowns of loads, as they are costly in time and risk (loss, theft, breakage, break in the cold chain).

– Settlement areas. In addition to certain financial aspects such as the cost of building construction or purchase and taxes (or rather non-taxes in urban free zones[6]), the logistician takes particular interest in the means of access: highways and road widths for land vehicles, access to rail, sea or air.

– Surface of the land and the depot. If the size of the depot depends on the number and size of the references to be stored or transported, the configuration of the building also depends on the constructible surface of the land. Thus, in view of this constructible surface, the depot will be built on a single plane or on several planes (floors): the material is not the same (e.g. elevators, robots and shuttles of storage, shelving by accumulation, etc.) (see Sheets 17 to 22 for the management of a warehouse in its entirety).

5 As a reminder, more and more decrees require companies to reprocess certain end-of-life products such as electrical waste and electronic equipment or furniture.

6 The urban free zone is a territory where companies can have tax exemptions (source: https://www.service-public.fr/professionnels-entreprises/vosdroits/F31149).

The barycenter is a calculation method that lets you find out the center of gravity between several points. This scientific method is widely used in logistics to calculate the ideal location (in an orthonormal benchmark) of a logistics warehouse or distribution center.

The barycenter method consists in taking all the geographical coordinates of the sites to be served into account (abscissa and ordinate), which are weighted by the share of traffic (volume in tons). The barycenter is the orthonormal reference point calculated by doing:

– for the abscissa, the sum of each abscissa multiplied by each ton of the serviced site divided by the total sum of the tons in volume of all the serviced sites;

– for the ordinate, the sum of each ordinate multiplied by each ton of the serviced site divided by the total sum of tons in volume for all the sites served.

However, the barycenter must be valued in view of the terrain constraints.

Box 6.4. *The barycenter (source: http://logistique-pour-tous.fr/calcul-barycentre-quelques-lignes/)*

6.2.3. *The alternative between warehouses and/or platforms*

After having determined the choice of construction and location, the alternative is to assess whether it is preferable to have a warehouse or a platform, knowing that a warehouse is a place of storage, in other words that the products remain on site for an average of more than 24 hours, while a platform does not have the virtue of storing products, it is a place of transit. In the platform, where products remain for less than 24 hours, we talk of cross-docking.

A warehouse or platform may be non-specialized (all products are found on site, but the specifications of the products must be taken into account when placing them in the warehouse, e.g. flammable products may be placed in closed spaces away from other products) or specialized by product (dry).

The region or departments served can also be chosen as the criterion of choice. For example, more often than not, a densely populated area has warehouses to respond very quickly to any demand.

Lastly, it should be noted that a warehouse or platform can be found anywhere, in the city, on the outskirts of the city or far from the city. In doing so, large warehouses are usually located far from city centers in order to benefit from the financial advantages and larger building space. As for the hub, which is a transit point, it is often located on the outskirts of cities. When the platform is located on the outskirts or in the city, it is referred to as an urban distribution center.

6.2.4. *The case of the urban distribution center (UDC)*

An urban distribution center (UDC) is a platform often located a few kilometers away from the city center. It is a place of grouping and unbundling whose primary purpose is the management of flows to dense areas.

The UDCs were mainly set up in response to the ban on large trucks entering the city center. In fact, the mayor – basing himself on Article L2213 of the *code général des activités* (French general code of activities) – can *"apply by-laws to prohibit, at certain times, the access of certain lanes of the agglomeration or to reserve this access to various categories of users or vehicles, to regulate the stopping and parking of vehicles or to reserve places to facilitate the stopping of vehicles carrying out a loading or unloading of goods"* (source: Cerema [CER 14])[7]. In addition, the UDC is a response to the introduction of cleaner routes, through the use of the so-called green modes of transport, that is to say little or no pollutants (LPG, hybrid, electricity).

Since January 1 2017, the possibilities of access to Grenoble's extended city center have been modified for goods vehicles. The objective of this change is twofold: to promote the efficiency of urban logistics and improve air quality in the Grenoble metropolis.

PROMOTE SUSTAINABLE URBAN LOGISTICS AT THE METROPOLIS SCALE

Urban logistics concerns all goods entering and leaving the Grenoble metropolis (transport of goods from private sector companies, waste collection, postal services, but also transport of goods by private individuals as part of their purchases). In our

7 http://www.territoires-ville.cerema.fr/reglementation-locale-du-transport-de-marchandises-a696.html

territory, it represents 24% of the total kilometers travelled by all vehicles, 36% of the fine particles emitted and 46% of nitrogen oxide emissions.

To meet the challenges of health, economic dynamism and development raised by freight transport, some twenty partners (including the City of Grenoble, Grenoble-Alpes Métropole and SMTC) signed an action plan for sustainable urban logistics in our metropolis on January 29 2015. The latter included 16 actions, including changes in delivery regulations, improvement of the service to the city center via an Urban Distribution Center (CDU), and optimization of the operation of delivery areas.

THE EXPANDED TOWN CENTER: AN RTA FOR GOODS VEHICLES

In order to limit the rate of fine particles emitted by freight transport vehicles, the Metropolis and the SMTC set up a Restricted Traffic Zone (RTA) on January 1 2017. This area, now prohibited to the most polluting heavy-duty vehicles and light commercial vehicles from Monday to Friday from 6am to 7pm, affects the enlarged city center: the Grenoble territory north of the main boulevards.

Box 6.5. *More responsible urban logistics for more breathable air (source: extract from http://unevillepourtous.fr/2017/01/04/une-logistique-urbaine-plus-responsable-pour-un-air-plus-respirable/ – January 4, 2017)*

6.3. Sheet 35: Urban Logistics or Last Mile Logistics

Key point

Urban logistics or last mile logistics is defined as *"the art of transporting the flows of goods that enter, leave and circulate in the city in the best conditions"* [PAT 02].

6.3.1. *The problems of urban logistics*

The issue of population density concurs with the issue of urban logistics. In 2012, the GCdD (*Groupe Caisse des Dépôts*) noted that 80% of the French population lived in the city and was responsible for 75% of greenhouse gas emissions, and that this figure would only increase as France is expected to have an additional 2.5 million inhabitants by 2030. This population growth increases deliveries (e-commerce, home delivery[8], shops) and the associated inconveniences such as traffic jams, noise and pollution. The logistician is therefore faced with a multitude of operators and urban spaces that must be managed with the best balance between cost, quality and time.

6.3.2. *Operators in urban logistics*

When we talk about logistics in general, three types of operators should be considered: the producer, the distributor and the end consumer (see Sheet 8). However, when we look at urban logistics, two other categories of operators must be added, namely institutional operators (State, municipality) and the urban-dweller (see Sheet 12).

The weight of institutional operators and urban-dwellers plays an increasingly important role in the act of delivery and, therefore, in the logistics business. Thus, in order to reduce noise, traffic jams and pollution, several orders can be made by institutional operators (see Article L2213 of the *code général des activités* – The French general code of practice) in order to limit the type of vehicles entering the city center. Delivery times can be decided upon as well. These constraints change the way delivery is handled at this part of the supply chain. One of the answers is, for example, the setting up of urban distribution centers (see Sheet 34).

6.3.3. *The division of urban logistics spaces*

To make it easier to track deliveries, the urban perimeter is divided into several zones. In 2006, Boudouin listed the four zones of these urban logistics spaces (ULS), starting from the furthest away and going towards the city center:

– Urban logistics zones (ULZ) are located within the outskirts of urban areas and include several transport companies. These areas are generally not

8 Bardin and Moesch [BAR 13] speak of purchase delivery, defined as *"a delivery service at a point x (relay point, home, office, hotel...), in a territorial logic, following a person-to-person purchase (commercial-customer), organized by a dedicated and pooled structure, coordinating multiple operators of a defined perimeter. The scope of purchase may differ from the scope of delivery"* (source: CRET-LOG – Cluster paca Logistique).

affected by driving permit requirements (vehicle weights, schedules) and, therefore, neither are delivery and collection practices [GON 13].

– Urban distribution centers (UDC) are usually located a few kilometers away from the town center. Grouping and unbundling platforms manage flows to dense areas. For example, heavy-duty trucks (over 3.5 tons) unload the goods that are then transferred to smaller vehicles going to different districts in the city (see Sheet 34).

– Thirdly, within neighborhoods and streets, the vehicle reception points (VRP) are infrastructures in the form of micro-logistics platforms, which are dedicated to allocating part of the road network to parking vehicles for the transport of goods. There are three sub-categories of infrastructure at this level, namely (1) local logistic spaces or local delivery areas (LDA), (2) goods reception points (GRP) and (3) delivery areas.

– Lastly, urban logistics boxes (ULB), which are located in streets or buildings. These are small structures – mobile or fixed – that provide an interface and a place of temporary storage of goods, in order to better optimize the transition of delivery rounds.

6.3.4. *Means to limit the consequent and polluting transport of goods*

The limitation of the numbered transport and pollutant of goods is a challenge for the logistician. On this issue, several reflections and actions have been taken. These include the following:

– The setting up of urban distribution centers, which aim to unburden the large tonnage of trucks and transfer them to vehicles that are less than 3.5 tons. Admittedly, some trials have not been conclusive [GON 13], but it is nonetheless a highly desirable and tested approach.

– The massing and pooling[9] of goods within the means of transport, including on the way out and back, in order to avoid empty deliveries.

9 Massing is the grouping of consignments of goods before they are shipped. As for pooling, it consists of sharing the resources of several services and/or companies in order to maximize the use of these resources. Massing and pooling are particularly well thought out in terms of delivery, subsequently using means of transport because any means of transport used minimally or to depletion (often in returns) is an important cost for the company.

– The use of low-emission or non-polluting vehicles, which includes electricity, LPG hybrid or even bicycles.

– The physical Internet, which was recommended by Eric Ballot and Benoit Montreuil [BAL 14], *"The physical Internet is a global logistics system built on the interconnection of logistics networks, through a standardized set of collaboration protocols, modular containers and intelligent interfaces for increased efficiency and durability"*. The physical Internet is valued as a *"hyper-connected global logistics system"* that seeks to pool all the global networks to optimize the delivery of goods.

– Drones and even some aircrafts are increasingly being considered/engineered in the delivery of goods [MOR 15].

– 3D printers[10], which even if the cost is still quite substantial and the interest in mass producing at low revenue does not yet make 3D technology popular with manufacturers, is a technology that is taking off and finding an important place in urban logistics [FER 17]. Thus, the 3D printer makes it possible to offer a product "on site" within very short lead times and reduces transport costs, especially upstream of the process.

6.3.5. *Reverse logistics*

Although reverse logistics includes the entire supply chain from downstream to upstream, we place it in urban logistics because that is where it finds its supply. It is in fact from the finished product of the delivered product (or in delivery) that this logistics begins.

Several terms are used to describe reverse logistics: reverse distribution [CAR 98] return logistics [BEA 99], reverse logistics [ROG 99] or inverse logistics [LAM 03]. This being said, the reverse logistics term predominates. For example, Rogers and Tibben-Lembke [ROG 99] define reverse logistics as *"the process of planning, implementing and controlling, in a rational and cost-effective manner, the flow of raw materials, semi-finished products,*

10 In May 1980, the Japanese Hideo Kodama filed a patent on the Rapid Prototyping (RP) technology that is the basis of today's 3D printer. In 1983, the French Jean-Claude André, Olivier de Witte and Alain le Méhauté took the initiative of stereolithography technology, but due to a lack of resources, development did not continue. On August 8, 1984, Chuck Hull, an American, filed his patent and was recognized as the instigator of the 3D system (Source: https://www.sculpteo.com/blog/2017/03/01/whos-behind-the-three-main-3d-printing-technologies/).

finished products and related information from the point of consumption to the point of origin, with the purpose of recovering or creating value or improving waste disposal".

Some authors distinguish green logistics from reverse logistics [ROG 01]. However, ultimately, for Lambert and Riopel [LAM 03], inverse logistics must combine reverse distribution and green logistics with the aim of recycling, refurbishing, waste management and container reuse, and be assessed as follows:

– reverse distribution involves replacement products, warranty repairs and product returns;

– green logistics seeks to substitute and reuse materials, reduce sources, save resources, design for the environment, improve productivity and eliminate waste.

In 2004, Lambert and Riopel set out three reasons for the return of the finished product from downstream to upstream:

– First, commercial reasons, whose main purpose is to increase the level of customer satisfaction. It may contain (1) technical (or after-sales service) problems, including repair (under warranty or not), refurbishment and recall campaign and (2) trade repossessions related to excess stock and shipping error.

– Second, legal reasons, which refer to the increase in legislation requiring a reduction in the number of landfilled or ultimate waste. This refers to the recycling of end-of-life products, packaging materials and production waste.

– Third, economic reasons, for reuse (packaging material, reusable containers, rental products) and recovery (production waste).

Within the framework of reverse logistics, two policies are sought by the logistician. First, a reduction in the amount of ultimate waste; in other words, waste that is landfilled, and therefore, not reprocessed for return to the market. Then, the technique of re-production or remanufacturing involves taking a used product and giving it a second life, by integrating new components [MIC 10, ELK 10].

I – Waste producers, in addition to the waste prevention measures they take, and waste holders shall organize the management of waste in accordance with the hierarchy of treatment methods, defined in II, 2° of Article L. 541-1.

The order of priority of the treatment method may be particularly changed for certain types of waste if provided for in a plan established under Articles L. 541-11-1, L. 541-13, L. 541-14 and L. 541-14-1, covering the territory where the waste is generated.

This order of priority may also be changed if justified in the light of environmental and human health implications, in addition to technical and economic conditions. The person producing or holding the waste shall then make the necessary supporting documents available to the competent authority.

II – Manufacturers or waste holders may only dispose of, or eradicate, in waste storage facilities as ultimate waste.

For the purposes of this article, ultimate waste is waste which is no longer likely to be reused or recovered under the technical and economic conditions of that time, in particular by extraction of the recoverable part, or by reducing its polluting or dangerous aspect.

III – The terms of this article shall not apply to households.

Remanufacturing: European remanufacturing network (source: https://www.remanufacturing.eu/)

The ERN was created for 2020 by the European Commission, with the aim of *"facilitating the development of sustainable partnerships between re-production organizations"*. The consortium consists of six universities: Strathclyde in the UK, Fraunhofer in Germany, Linköping in Sweden; VTT Technical Research Center in Finland, Grenoble INP in France and Delft University of Technology and Circle Economy in the Netherlands.

Box 6.6. *Ultimate waste and re-production (remanufacturing) Ultimate waste: Article L541-2-1 of the French LAW n° 2015-992 of August 17, 2015 - art. 70 (source: https://www.legifrance.gouv.fr)*

7

The "Companions" of Logistics

7.1. Sheet 36: Traceability

Key point

Traceability was created to respond to health crises and counterfeiting. Its main purpose is to establish and facilitate a relationship of trust with the end consumer, through the presence of significant regulatory and legal constraints. ISO 8402:1994 defines traceability as *"the ability to retrieve the history, use or location of similar item(s) or activity(s) by means of a registered identification"*. Traceability is sometimes seen as a tool, as a component of Supply Chain Management, or as a management approach in its own right [KAR 10, LAZ 14].

7.1.1. *The challenges of traceability*

The implementation of traceability is a challenge for both the consumer and the company:

– for the consumer, it facilitates the relationship of trust through technical, regulatory and legal constraints. For example, every company must keep certain documents – for a period of 3 to 10 years – in order to prove its good faith in the event of a claim; however, above all, it must be equipped with all the tools needed to monitor a product and its components in the event of a recall for health reasons;

– for the company, it is a customer loyalty tool thanks to its mastery of its supply chain and its ability to react quickly in any needed case. Traceability for the company therefore requires extensive coordination between internal operators (all staff) and external operators (companies, traders) in the supply chain.

7.1.2. *Traceability functions: tracking and tracing*

Two functions are associated with traceability: tracking and tracing. The tracking function *"refers to the issue of location (finding an entity)"* whereas the tracing function is used *"to monitor, in order to establish the trace"* [FAB 06].

The implementation of tracking requires the possession of systems and tools to identify, observe and analyze information. To date, there are three types of codification systems: barcode, Data Matrix code and RFID tag. Monitoring tools can range from paper-based tracking to specialized software packages. As for the tracing function, the path the product has followed is traced. The tracing function requires a good tracking system.

7.1.3. *Types of traceability*

There are two types of traceability: upstream/internal/downstream and ascending/descending:

– upstream/internal/downstream traceability refers to the "horizontal" representation of traceability. Upstream traceability refers to the part of traceability before an operator in the supply chain becomes responsible for product monitoring. Internal traceability corresponds to the follow-up of the product by a particular operator. And downstream traceability concerns tracking the product when it is transferred to a third party;

– ascending and descending traceability's purpose is to find a product or component. For example, Product 1 is composed of 75% of Component A and 25% of Component B, and Product 2 is made up of 10% of Component B and 90% of Component C. In the event of an alert on Component A, it is possible to descend to Product 1, which may be a "victim" of the identified anomaly (descending traceability). In the event of an alert on Product 2, it is possible to ascend back to Components B and C, which may be responsible for the identified anomaly (ascending traceability). Note that in this example, if Component B is suspected, it will also be necessary for it to go down to Product 1, which may also be "contaminated".

7.1.4. *The four levels of traceability tools*

Pellaton and Viruega [PEL 07] considered four levels of (IT) tools used in traceability:

– a first level that supports information through barcodes and RFID tags;

– a second level that seizes information via readers and devices;

– a third level that stores and allows information retrieval through software packages;

– a fourth level that reflects the inter-organizational aspect expected by traceability and whose role is the sharing of information via EDI (Electronic Data Interchange, see Sheet 38).

7.1.4.1. *Barcode, Data Matrix code and RFID tag*

Lazzeri [LAZ 14] provided an overview of the three main tools used in traceability, from which some key elements can be extracted:

– the barcode or barcode ID *"consists of a sequence of black and white vertical bars containing information encoded according to the rules that determine, for each character, the position and size of the bars. Linear scanning barcodes are described as one-dimensional. Reading is done by scanning and then the information is transferred to the database"*. The capacity is from 1 to 40 characters and its unit cost (price as an indication) is from 0.01 to 2 euros. Its main quality is its ease of use and its weak points are (1) low reading distance, (2) the requirement for high print quality, (3) limited memory capacity, (4) a non-writable label, (5) stored information sensitive to the environment, such as shocks and dirt, (6) reading it facing the optical reader, and (7) reading one by one;

– the Data Matrix code is a *"multiple scan or image barcode. They are called two-dimensional. The Data Matrix is a matrix, 2D or two-dimensional barcode in the form of a square or rectangular symbol, consisting of dots or squares juxtaposed together. This representation is an ordered grid of black 'dots' and white 'dots', bounded by motives. Depending on the implementation constraints of the Data Matrix (media configuration, available space on product, nature and volume of the data to be encoded, printing process...), it is possible to print the Data Matrix, either square or rectangle"*. The capacity is from 5 to 3000 characters and its unit cost (price as an indication) is from 0.01 to 2 euros. Its main quality is its high capacity of use and its weak points are (1) a non-writable label, (2) the need for a specific decoder reader, (3) stored information sensitive to the environment, such as shocks and dirt, and (4) a one-by-one reading;

– the RFID tag for Radio Frequency Identification *"is a technology that uses radio frequency to exchange data between an antenna and an object (product, document, person, etc.) in order to identify, locate, categorize and trace it"* [MEV 09]. The RFID tag consists of an electronic tag and an antenna. The capacity is a few characters with several Kbytes and its unit cost (price as an indication) is from 0.5 to 20 euros. Its main qualities are its great usability, its remote/volume and blind reader, unrequired human intervention, a possible rewriting (addition or modification of information) and a possible password protection. Its weak points are (1) financial (high cost), (2) technological (poor operation when close to metal and humidity, system compatibility, differences in high frequency regulation between Europe, Asia and the United States), and (3) ethical (uncertainties regarding health risks of radio waves, jobs, individual freedom).

7.1.4.2. *Product coding*

Product coding allows information that facilitates product identification and tracing to be held. The information found in coding includes the country of origin, manufacturer, purchase and sale price, package size, number of items/packages or sub-packages[1].

GS1 France[2] sets international standards for the marking of consumer, logistics and shipping units:

– for consumer units (CUs), the code used is GTIN-13: *"the GTIN code is a code identifying the products, services or packages, composed of numerical characters (company prefix, article code, control key). It is translated by the symbology of the 'bars' into an EAN-13 symbol"*. The GTIN-13 code is represented by a series of 13 digits, broken down into a company prefix (between 6 and 11 digits), a product code (from 1 to 6 digits) and a control key. The code formats are EAN-13 for European Article Numbering, which is the most widely used, EAN-8 for small

1 In terms of conditioning, the term uses abbreviations of three types. The CSU or Consumer Sales Unit specifies the number of items, the PCB identifies the number of CSU per package (knowing that PCB is translated by *Point de Couverture Brute* or more generally by ComBien – Gross Coverage Point) and the SPCB considers the number of products per sub-package.

2 http://www.gs1.fr/Notre-offre/Supply-Chain-Interconnectee/Codification/Les-standards-pour-identifier-vos-unites.

products, GS1 Data Bar, GS1 Data Matrix and the GLN code for Global Location Number, in order to identify a place of work[3];

– for logistics units (LU) which are cartons, pallets, packages, etc., two types of identification can be chosen, namely GTN-13, which translates as an EAN-13 barcode on the carton, or *"GTIN-14 [which] is symbolized by the ITF-14 barcode (for carton printing, used for in-store reading) or the GS1-128 barcode to add further traceability or logistics information"*;

– for shipments. GS1 France adds that a pallet must be identified by an SSCC code: *"Pallets must be identified by a Serial Shipping Container Code (SSCC), translated into a GS1-128 bar code. The SSCC is a code allowing each shipping unit to be uniquely identified with additional information (lot number, date, pallet number, size, weight, etc.). It is used for the automation of shipping, delivery and product processing, and processes in EDI messages. The creator of the shipping unit is responsible for assigning the SSCC (manufacturer, wholesaler, logistics provider, carrier, distributor, etc.)"*.

7.2. Sheet 37: Freight Transport

Key point

The *Ministère de l'Environnement, de l'Energie et de la Mer* (Ministry of the Environment, Energy and the Sea) (2017)[4] recalls that transport is *"the activity which consists of moving a physical good or a person from one point to another, with a purpose other than the movement itself. The transmission of electricity and information in particular are excluded from this field. A home-work trip is transport, since it is intended to make a person available at work. On the other hand, a trip by bicycle with return to the starting point is not transport, because its primary motive is leisure time provided by this activity"*. Sheet 11 presents the different types and operators of freight transport. Transport management is also associated with the use of the InCoTerms, as stated in Sheet 25. As for the transported goods, their follow-up is facilitated by the tracking tools mentioned in Sheet 36. However, other aspects need to be considered in the management of freight transport.

3 http://www.gs1.fr/Notre-offre/Produit-Digitalise/Des-codes-et-des-barres/Le-code-a-barres.
4 MINISTÈRE DE L'ENVIRONNEMENT, DE L'ENERGIE ET DE LA MER, *Chiffres clés du transport*, Service de l'Observation et des Statistiques, February 2017.

7.2.1. *The problem with freight transport*

In 2017, the *Ministère de l'Environnement, de l'Energie et de la Mer* (Ministry of the Environment, Energy and the Sea) displayed the latest freight transport data from 2015. With regard to the land transport of goods (excluding oil pipelines), road haulage continues to dominate. Of the 323.1 billion tons per km transported, road haulage accounts for 87.09% (or 281.4 billion tons per km), rail for 10.58% (or 32.2 billion tons per km) and inland waterway for 2.33% (or 7.5 billion tons per km). For air transport, 2.2 million tons of goods are shipped from French airports, usually to international destinations. French maritime transport remains stable at 350 million tons in 2015 (see Sheet 11, section 1.7.1 for an overview of the advantages and disadvantages of each mode of transport).

The main mode of freight transport, French road transport, is particularly affected by competition from the New Member States (NMS) of the European Union: Bulgaria, Croatia, Estonia, Lithuania, Poland, Czech Republic, Romania, Slovakia and Estonia. The main reason for this is that the wage costs of the NMS are lower than the French costs.

7.2.2. *Regulations in terms of international road haulage and cabotage*

Since January 18, 2017, the *Ministère de la Transition Ecologique et Solidaire* (Ministry of Ecological and Solidarity Transition) has posted the rules, in terms of international road freight transport[5], on its website. Among the regulations cited are the following:

– the newsletter dated June 21, 2010, which refers to Decree no. 2010-389 of April 19, 2010 and which has been included since January 1, 2017 in articles of the transport code. This newsletter concerns road freight cabotage[6]. It states that *"cabotage operations are limited for the same lorry*

5 https://www.ecologique-solidaire.gouv.fr/transport-international-marchandises-et-cabotage.
6 *"A road haulage cabotage operation means any transport of goods (loading, unloading) between two points in the national territory, carried out by a non-resident. Road cabotage of goods may be carried out, subject to conditions, on French territory by an undertaking established in a Member State of the European Union or the European Economic Area. The aim of the scheme is to allow cabotage to become an extension of international transport, in order to avoid unladen journeys"* (source: https://www.ecologique-solidaire.gouv.fr).

or, if it concerns a combination of vehicles, for the same road tractor: three operations within a maximum period of seven days from the complete unloading of the goods which have been the subject of the prior international transport, where such transport is destined for France; a single cabotage operation within three days of the empty vehicle’s entry into France, where the prior international road transport is not intended for France. This operation must also be completed within seven days of the unloading in international transport”;

– Act no. 82-1153 of December 30, 1982 and Article R. 3411-13 et seq. which lists the documents that every vehicle must hold. These documents are (1) the administrative title of transport (that is, driving license), (2) the consignment note[7], (3) the rental document if necessary, (4) the driver’s attestation for international transport, and (5) any document relating to international cabotage[8].

7.2.3. *Types of road transport and loading*

Two types of transport are present, combined or not:

– own-account transport: here, the company owns its own freight transport vehicles;

– third-party transport: the company does not own goods vehicles and uses external service providers.

---

7 *“The consignment note is a contractual document that harmonizes the general conditions of international land transport and deals with the carrier’s liability. This letter is sometimes referred to as the CMR consignment note because it is covered by the* Convention Relative au Transport International de Marchandises par la Route *(Convention on the International Carriage of Goods by Road), abbreviated as CMR. Ratified in Geneva in 1956 and put into effect in 1958, this convention requires a consignment note to be drawn up if the country of departure or the country of destination of the goods, or both, have signed it. More than 50 countries have ratified this convention, including all European Union member states”* (source: http://www.journaldunet.com/management/pratique/vie-professionnelle/18575/la-lettre-de-voiture-cmr.html).

8 https://www.legifrance.gouv.fr/affichCodeArticle.do;jsessionid=8E57CFBC6F4F01A5E8BCD8091B80D993.tpdila18v_2?idArticle=LEGIARTI000033450565&cidTexte=LEGITEXT000023086525&dateTexte=20170113.

Two types of loading can be proposed:

– Less Than Truckload (LTL): the vehicle is not fully loaded;

– Truck Load (TL) or Full Truck Load (FTL): the truck trailer is fully used.

7.2.4. *Types of road vehicles*

For road freight transport, there are four types of rolling stock, namely:

– carriers or individual vehicles which are vehicles in one piece;

– road tractors which correspond to the motor vehicle and where the goods are transported in a semi-trailer attached to it;

– semi-trailers which are not motorized and where the goods are placed. Semi-trailers only have wheels at the rear. They attach to road tractors to form articulated vehicles. When the articulated vehicle is attached to a semi-trailer, it is called a double road train;

– trailers are also motor-less vehicles in which the goods to be transported are placed. Unlike semi-trailers, trailers have front and rear axles. A road train is a road tractor vehicle to which a trailer or semi-trailer is coupled, with its front end resting on an undercarriage.

Article R. 311-1 of the highway code[9] lists the different categories of vehicles that can circulate in France. As for Article R. 433-1 of the highway code, it specifies the conditions for authorizing traffic for vehicles *"of an exceptional nature due to their size or mass exceeding the statutory limits"*[10]. The elements relating to the transport of large tons are therefore specified in this article.

Directive 2007/46/EC of the European Parliament and of the Council of September 5, 2007 establishes a *"framework for the approval of motor vehicles, their trailers and systems, components and separate technical units intended for such vehicles"*. Transposition texts specifying the construction

9 https://www.legifrance.gouv.fr/affichCodeArticle.do?cidTexte=LEGITEXT000006074228&idArticle=LEGIARTI000006841575.

10 https://www.legifrance.gouv.fr/affichCodeArticle.do?cidTexte=LEGITEXT000006074228&idArticle=LEGIARTI000006842354&dateTexte=&categorieLien=cid.

elements of a vehicle are available on the website www.legifrance.gouv.fr in conjunction with this Directive.

Within the environmental framework, heavy-duty vehicles must meet anti-pollution standards, the latest being the Euro VI standard.

The Euro Standard was introduced by the European Union in 1988 for heavy-duty vehicles (Euro 0 to VI) in order to limit emissions of pollutants from road transport.

It sets increasingly constraining standards for manufacturers, who are obliged to place less polluting vehicles on the market.

Thus, all new vehicles must now comply with the Euro VI standard, pursuant to Regulation no. 595/2009 of the European Parliament and the Council of June 18, 2009.

The currently applicable Euro VI standard makes it possible to significantly reduce the limits on pollutant emissions from heavy goods vehicles, compared to the Euro III standard. For example, the fixed limits for nitrogen oxides and particles were both reduced by 92% between 2001 and 2014.

Compared to the Euro V standard, the Euro VI standard requires a further reduction in limit values of 80% for nitrogen oxides, 50% for particles and more than 70% for hydrocarbons.

In 23 years (between Euro 0 and Euro VI), NOx (nitrogen oxides) will have been divided by 36, HC (hydrocarbons) by 18 and particles by 35.

It should be noted that research on Euro standards has led to the development of technological advances in engines that have made it possible to reduce the average fuel consumption of vehicles.

Thus, as a result of stricter standards, the overall quantities of pollutants emitted have been considerably reduced.

Box 7.1. *The Euro VI pollution emission standard for heavy-duty vehicles – Clean vehicles (source: extract from https://www.ecologique-solidaire.gouv.fr/normes-euros-demissions-polluants-vehicules-lourds-vehicules-propres - 12/01/2017)*

7.2.5. *Other types of freight transport*

Even though the name of the contract for road haulage is the consignment note, it takes on a different name for other types of transport. For example, for air transportation, the contract is referred to as an air waybill (AWB), and for sea transportation, the document is called the bill of lading.

Generally speaking, containers are used for shipping, railways and even air transport. Container loading can be complete (Full Container Load) or partial (Less than Container Load). When the loading is partial in the maritime context, the term maritime grouping is used.

We talk about packing and unloading in the loading and stowing of the container and vice versa. Each container must comply with safety regulations (see Box 7.1). We refer to carrier haulage when the shipping carrier takes the empty container to the loader/shipper and brings it to the port of departure full. The opposite technique is merchant haulage, that is, the loader/shipper takes the container to its storage area, packs it and transports it to its destination.

Packing consists of loading and stowing the goods in the container. It complies with common sense rules and must particularly take mass–volume compensation, mass balancing and a good center position of gravity into account. Incorrect packing can have consequences on the goods, the container itself, persons or damage reports. The opposite operation is removal.

The characteristics of each container are indicated on its CSC plate (1972 International Convention on the Safety of Containers). The door on the right has a larger reminder, but only the indications of the CSC plate are valid.

Containers are designed to withstand these weight limits. There is a risk of accidents due to excess and ISO standards do not allow any additional kilograms. Agents of the shipping company are responsible for any container found to be overweight and delivered to the row. The total weight of a container (gross weight) is the sum of the weight of the empty container (tare) and the weight of the goods (payload).

Since July 2016, the revised Solas Convention *has required the weighing of containers before loading on a ship and the advanced declaration by the shipper of the verified gross mass (VGM) to the ship-owner. The application of this rule still gives rise to differences of interpretation from one country to another [...].*

Whenever containers are outside a private depot, whether full or empty, on a terminal, aboard a ship, a carriage or a wagon, they shall be sealed. It must be installed on the left side of the right-hand door of the container, which is reputed to be impossible to dismantle. Seals for full containers are made up of a hardened steel rod and a metal cylinder socket covered with a transparent plastic protection, allowing the marking to appear. The purpose of this protection is to prevent tampering with the marking. Metallic seals with blue or white cylinders are applied to full containers and fluorescent orange plastic seals are applied to empty containers.

The transport of dangerous goods by container is subject to special packing and reporting rules governed by the IMDG (International Maritime Dangerous Goods) code.

Box 7.2. *Container packing and safety standards (source: http://www.lantenne.com/Empotage-des-conteneurs-et-normes-de-securite_a32691.html)*

7.3. Sheet 38: Information and Communication Technologies, and Information Systems

Key point

The management of a growing number of material/product references is becoming increasingly important. In this context, the use of Information and Communication Technologies (ICT) has become a prerequisite in logistics. ICTs are defined as *"the technical components that enterprises purchase, develop and combine to form the technological infrastructure that will enable their information systems (IS) to function"* [LAU 10]. In the book published by Agora Fonctions, *La Supply Chain : Une fonction stratégique en action* [AGO 11], several logistics managers are asked a question about the role and stakes of ICTs. The results show that ICTs allow (1) tracking in terms of traceability, (2) fraud management, (3) making exchanges more reliable in real time, (4) customer service, (5) invoicing, (6) business to consumer activity or e-commerce, (7) control of leeway, (8) management of waste and dormant stock, (9) the optimization of industrial capacities, (10) the compatibility or interoperability of technologies, (11) management of the last mile, often associated with the concept of urban logistics, (12) the Research and Development process, and (12) the possibility of giving access to external operators on the follow-up of their orders.

7.3.1. *The difference between a software package and a software*

It is not always easy to know if you are dealing with a software package or software. The definitions resulting from the Decree of December 22, 1981[11] for these two terms inform that:

– a software package is a *"complete and documented set of programs designed to be provided to multiple users for the same application or function"*, the foreign equivalent of which is a *package*. By definition, the software package is therefore opposed to specific developments;

– a software is a *"set of programs, procedures and rules, and possibly documentation, relating to the operation of a data processing set"*.

A software package can be considered as a "standard" program, that is, neither enriched nor customized to respond to a set of operators; in addition, a software package can be "customized" to the specific needs of the user. When too many changes are made, this usually leads to a problem of interoperability between intra- and inter-enterprise machines.

7.3.2. *ICT families in logistics*

Many ICTs are used in logistics management. Evrard Samuel *et al.* [EVR 11] cited two main families of logistics implements, namely:

– the SCP or Supply Chain Planning. With the help of various information (order quantities, production schedules, etc.), SCP allows you to establish forecasts based on current situations. This family includes, for example, APS (Advanced Planning Scheduling), CRM (Customer Relationship Management) and the TMS (Transport Management System);

– the SCE or Supply Chain Execution, also referred to as the LES or the Logistic Execution System. This family of implements manages the total order processing cycle. Technologies include AOM (Advanced Order Management), CRM, TMS, MES (Manufacturing Execution System) and the WMS (Warehouse Management System).

11 http://ensmp.net/cstic/020112fascicule_internet.pdf, in the official newspaper of September 22, 2000 for these two terms.

The nature of some software packages classifies them into two types of families.

To these two families, we can add:

– the SCEM or Supply Chain Event Management. This constitutes an operational link between the SCE and the SCP, by inserting or consulting unforeseen or unplanned events in "standard" implements.

7.3.3. *Software packages and software in logistics*

Logistics software packages are numerous. Here is a list of the main technologies listed in alphabetical order:

– AOM: Advanced Order Management. Its role is the registration and management of orders;

– APS: Advanced Planning and Scheduling. A decision support implement in the sales forecasting process. It is intended for planning the company's flows of goods, information and finance;

– CRM[12]: Customer Relationship Management. In charge of sales management, a direct link between the customer and the company;

– EDI: Electronic Data Interchange. Defined as *"electronic data interchange (EDI), it is a technique that replaces the physical exchange of documents between companies (orders, invoices, delivery notes, etc.) by exchanges, in a standardized format, between computers connected by specialized links, or by a (private) value-added network (VAN). The data are structured according to international technical reference standards (e.g. Edifact). For example, faxing an order and then entering it by an operator in the supplier company is replaced by sending information to a computer in the supplier company capable of interpreting the order"* (INSEE, consulted in 2017 (source: https://www.insee.fr/fr/metadonnees/definition/C1026));

– ERP: Enterprise Resource Planning. Its use is in the management of repositories and transactions. It is responsible for all processing (transactions

12 The C in CRM is sometimes used in the Customer or Consumer sense. The difference is valued in the idea that the consumer is the individual who buys, whereas the customer integrates, in addition to the retailer or the person who buys for a company, with the purpose of resale (source: http://wiki.answers.com/Q/What_is_the_difference_between_a_consumer_and_a_customer).

and executions) of classical functions: Finance, Accounting, Commercial and Sales, Production and Material Management, and Sales Administration;

– MES: Manufacturing Execution System. Designates the systems that deliver relevant real-time information on the execution of production orders, with the aim of controlling them from their launch to the finished products and optimizing production activities;

– MRP: Material Requirement Planning. A production and supply management software that calculates net requirements at infinite capacity. It is a set of techniques that use nomenclatures, stock/requirements and the Master Production Schedule in order to calculate material requirements. It is primarily seen as a planning technique to establish and maintain correct due dates.

– SRM: Supplier Relationship Management. In charge of purchasing management, which is a direct link between the supplier and the company;

– TMS: Transport Management System. In charge of controlling the transport flows;

– WMS: Warehouse Management System. Manages the preparation, follow-up and execution of activities in warehouses;

– YMS: Yard Management System. Manages depots, docks, parks and car parks with a view to synchronization.

> *"Operational CRM includes customer-focused applications, such as implements for sales force automation, customer service and call center support, and the automation of certain marketing tasks. Analytical CRM includes applications that analyze customer data generated by operational CRM applications in order to provide information to improve business performance management"*. Also for Laudon *et al.* [LAU 10], CRM software is available on three levels: (1) sales, which is concerned with account management, lead management, order management, sales planning and the cost of sales; (2) marketing, which focuses on campaign management, promotion channel management, event management, market planning, marketing operations and marketing analytics; and (3) service, which includes service delivery, customer satisfaction management, return management, service planning, help desk and call center, and service analytics.

Box 7.3. *CRM (source: Laudon et al. [LAU 10])*

"ERP (Enterprise Resource Planning) [...] has become an essential element of all companies for years. A common data base, a very structuring 'process machine' that gives a framework, methods, implements so that everyone can work together, share the same language, the same rules of operation".

Box 7.4. *ERP (source: excerpt from Rognon [ROG 15, p. 88])*

MRP *stands for Material Resources Planning or Material Requirements Planning or Manufacturing Resources Planning. [...] As the forerunner of ERP (Enterprise Resource Planning), MRP is always the central part of a computerized production management system. As a mixture of software and database, MRP has the main role of allowing production planning based on personnel, raw material, machine and time resources, in relation to a requirement to date or stock requirement.*

A distinction is made between MRP (Materials Requirements Planning), which was formally instigated in the United States in 1964/65 and which represents a material requirements planning method, and its development, MRP2 or MRP II (Manufacturing Resources Planning). This broader model, which integrates the management of all company resources (consumables, in other words "materials and components", and renewable, that is, "machine and labor capacity"), is a resource control system that relies on sales forecasting and product nomenclatures, and operates as high-flow MRP (that is, production planning that is based on forecasts).

Box 7.5. *The different facets of MRP (source: https://fr.wikipedia.org/wiki/Planification_des_ressources_de_production)*

The coupling between the different software packages is part of the search for international standardization, which is adopted at the national level (see Box 7.6).

Efficient coupling of production in the supply chain is critical for industrial dynamics. Indeed, the industry has widely adopted ISA-95 in order to facilitate the interoperability of computer applications related to the production system.

The ***ISA-95*** *covers the various operational areas that support production (logistics, quality, maintenance) to ensure the continuity of processes through the applications concerned: ERP (Enterprise Resource Planning), control systems*

(SCADA, API/PLC, SNCC/DCS, etc.), MES (Manufacturing Execution Systems), and also LIMS (Laboratory Information Management Systems), WES (Warehouse Execution Systems), LES (Logistics Execution Systems), CMMS (Computerized Maintenance Management System)... The ISA-95 takes the interactions of business development with the operational area into account (PLM – Product Lifecycle Management).

IEC/IEC 62264-1:2013 – Integration of Enterprise Control Systems (source: https://www.iso.org/fr/standard/57308.html)

IEC 62264-1:2013 describes the management domain of manufacturing activities (level 3) and its activities, and the interface content and associated transactions within level 3, and between level 3 and level 4. Its objectives are to increase the uniformity and consistency of interface terminology and to reduce the risk, cost and errors associated with the implementation of these interfaces. IEC 62264-1 can be used to reduce the effort associated with implementing new product offerings. The aim is to obtain a company system and control systems that interact and integrate easily.

NF EN 62264 (source: https://www.boutique.afnor.org/recherche/resultats/mot/62264)

Five NF EN 62264 sub-standards deal with the integration of the company control system and they are the counterpart of IEC 62264.

Box 7.6. *The ISA-95 standard, the ISO/IEC 62264-1 standard, the NF EN 62264 ISA-95 standard*

7.4. Sheet 39: Quality Implements and Logistics

Key point

Very early on, the link between quality and logistics was shown [FOR 97]. This aspect is reflected in the various standards that logistics uses like the ISO 9000 standard, which is first and foremost a standard derived from the world of quality. Several methods and implements are used in logistics such as the WWWWW(H)W (or the Six Ws and one H) method and the Ishikawa diagram [LAM 99, LED 01]. We present the main implements used.

7.4.1. *Shewhart's control chart*

American physicist and statistician Walter Shewhart (1891–1967) was the engineer behind the principle of control charts. Based on a given analysis period (hour, day, week, month, etc.), the control chart consists of analyzing an action. The control chart is based on three values: an average value that must be maintained, a high value that is desirable to obtain in the future and a low value that must be avoided. The control chart makes it possible to quickly visualize any deviation and carry out a correction action if necessary.

On the abscissa, we put the selected period (day, week, month, etc.) and on the ordinate, we consider the measured action such as, for example, the customer satisfaction rate.

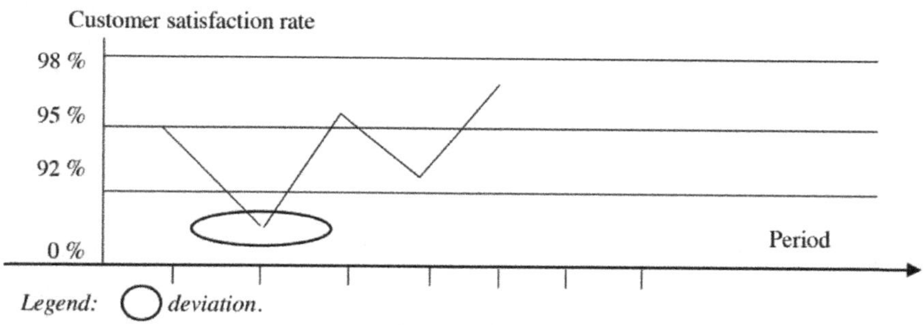

Figure 7.1. *Control chart*

7.4.2. *The Ishikawa diagram and the Causes and Effect Diagram with the Addition of Cards (CEDAC) of Fukuda*

Kaoru Ishikawa (1915–1989), a Japanese chemical engineer, is the designer of the diagram bearing his name, also known as a cause/effect diagram or fish bone diagram. This tool was designed in the 1950s and 1960s. It is usually structured around five main axes, called the 5M for Matter, Material, Method, Manpower and Medium. Each branch is subdivided into sub-branches in which the points to be dealt with are written, which may themselves be subdivided into sub-branches. The objective of this diagram is to try to visualize the actions to be taken in order to assess and/or improve a situation (see Figure 7.2).

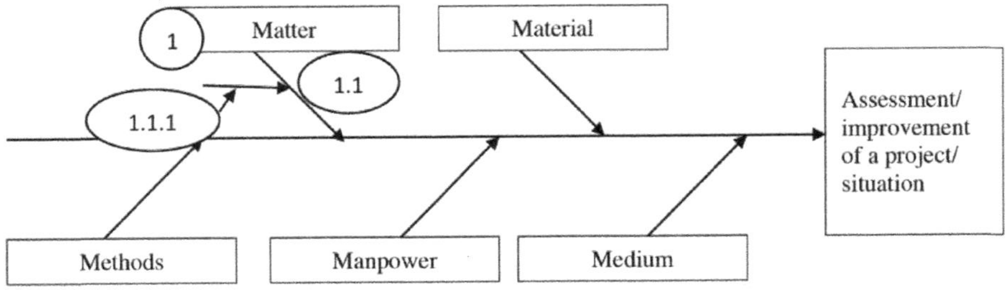

Figure 7.2. *The Ishikawa diagram*

For ease of reading, the Ishikawa diagram rarely shows more than two sub-branches. When each sub-branch is validated, a sheet is associated with each point to propose improvement actions. For example, if the "matter" axis refers to the "management of raw materials", that sub-branch 1.1 specifies an action on "evaluation tools", the sub-branch 1.1.1 will list all the techniques to be proposed as an implementation of (1) the Pareto or ABC technique (see Sheet 20), (2) a permanent inventory (see Sheet 20), etc.

The CEDAC (Causes and Effect Diagram with the Addition of Cards) diagram is a variation of the Ishikawa cause–effect diagram. It was developed by Ryuji Fukuda in the 1970s. It is based on a participatory method of operators with regard to resolving a problem[13]. Like the Ishikawa diagram, a graph is developed as shown in Figure 7.2. A working group is convened under the coordination of a leader. The project is defined over a resolution period, for example over two months, and meetings are scheduled weekly (or at other intervals) to deal with progress. The members of the group put post-it notes on the actions to be implemented. These post-it notes are positioned on the axes that have been previously defined as the most appropriate solution to the problem (examples of axes: materials, machines, personnel, methods, management, marketing, finance, performance indicators, maintenance, etc.). These post-it notes that state the actions to be taken will be of an identical color, in order to facilitate the follow-up. When the solutions are found and implemented, post-it notes (of a different color) are associated with these first post-it notes.

13 A participatory method of problem-solving is more commonly known as brainstorming.

7.4.3. *The Ademiecki–GANTT diagram and the PERT network*

Originally developed by the Polish engineer Karol Ademiecki (1866–1933) in 1896, the GANTT diagram was distributed by the engineer Henri L. Gantt (1861–1919) in 1910. The purpose of the GANTT diagram is to display the dates and achievements of a project or production on a single document.

The GANTT diagram is attached to the PERT network which analyzes the activities of a project, specifying the earliest and latest dates of each activity. The PERT network highlights the "critical path", in other words activities that have no leeway in their execution time. The PERT (Program Evaluation and Review Technique) method was developed by the U.S. Navy in the 1950s.

7.4.4. *The Pareto diagram by Joseph Juran*

The Pareto diagram is based on the Pareto or 20/80 method and its development in ABC, which is presented in Sheet 20. This calculation method classifies products into two or more categories depending on their importance. The more the product is classified in Category A, the more attention will be paid to it by the company. While the Pareto method is generally used in inventory management, it is perfectly applicable to other factors that are considered crucial, such as the classification of customers according to turnover. The Pareto diagram is therefore a graphical representation (often a bar graph) of the Pareto method. Joseph Juran (1904–2008) invented the Pareto diagram [JUR 99].

7.4.5. *Flow sheets*

Flow sheets are documents that present the steps of a process [LED 01]. The objectives of this type of document are to visualize the different stages of a process as precisely as possible, in order to identify and eliminate redundant tasks and bottlenecks, as well as to define improvement actions such as, for example, a computerization of a stage that is still carried out in paper format.

Purpose of the process studied: ____________														Date and author of the study:
Sheet: _x_/ _n pages	SERVICES												I.S. TYPES *	
Method: *Current *Future	SUPPLIER	R & D	Purchase	Production	Marketing	Finance	IT	Aft.er Sales Serv.	CLIENT	Distance (cm)**	Time (min)	Unit (part, weight)	Paper, Internet, EDI, ERP, etc.	Observations
Step-by-step description	O													
			O											
				O										
			O											
Intermediate Total														
Grand Total														
Current Grand Total difference- Proposed Grand Total														

**Legend: I.S. = Information System*

***It is important to use and clearly indicate the units of measurement used so that comparisons can be made. It is advisable to take the same unit for each step.*

Figure 7.3. *Example of a process flow sheet*

7.4.6. *Histogram*

The histogram consists of making a representation by columns and by classes of identical size. The obtained distribution often corresponds to a representation according to the standard normal distribution, in a "bell curve", also called the Gauss or Laplace–Gauss curve, named after the two mathematicians, astronomers and physicists (Pierre-Simon de Laplace: 1749–1827 and Carl Friedrich Gauss: 1777–1855) who developed it. This standard normal distribution tends to emphasize that all operating frequencies follow an "up and down" pattern.

7.4.7. *The WWWWHHW and the five Why's*

The WWWWHW or WWWWHHW is an acronym for Who? When? Where? What? How? (How much?) Why (P). It is a questioning method that is applied to a problem on the basis of these seven questions.

In questioning techniques, the five Why's is also a method used. For the problem being dealt with, the five why's method consists of asking the question "Why" five times in a row, in order to get as close as possible to understanding of problem.

7.4.8. *The PDCA method or the Deming–Shewhart wheel*

The authorship of Deming's wheel is attributed to the engineer Walter Shewhart (1891–1967), an American physicist and statistician, but it was William Deming (1900–1993) – an American statistician – who ensured its development in the 1950s.

The PDCA method for Plan (prepare) – Do (develop) – Check (control) – Act/Adjust is represented by a "virtuous circle" that is applied when we want to conduct a project or solve a problem (see Sheet 27).

7.5. Sheet 40: Environmental Decrees and Directives

Key point

With the desire to protect the environment, many decrees and Ecotaxes have been issued. These influence the logistics provider's business, through considerations based on the re-production and reduction of ultimate waste in product design (see Sheet 35), or even the obligation to call on third parties to reprocess products at the end of their life cycle.

7.5.1. *Ecotaxes*

An Ecotax is defined as *"a tax levied on a good, service or activity because of the damage it is likely to cause to the environment"* (source: *Journal Officiel* from February 1, 2011). There are many Ecotaxes affecting logistics. These include some of the best known:

– Ecotax or tax on heavy goods vehicles, which was due to come into effect on January 1, 2014 and which was suspended in October 2014 by the *Ministre de l'Écologie Ségolène Royal* (Minister of Ecology, Ségolène Royal) in the face of the revolt of French haulers;

– the tax on fuels or *TICPE* (*Taxe Intérieure de Consommation sur les Produits Energétiques* – Domestic Tax of Consumption on Energy Products) which can be assimilated to Ecotax. Road haulers of goods are entitled to a partial refund of this tax. Since July 1, 2017, the *Service National Douanier de la Fiscalité Routière (SNDFR)* (National Customs Service for Road Taxation) has been the single point of contact for this reimbursement[14];

– the eco-participation or eco-contribution is a tax paid by the purchaser of an electrical, electronic or furniture appliance, or even a mobile good. This is to treat the collection and treatment of waste. The introduction of this Ecotax forces companies to take back the goods with a view of recycling, either by themselves or by an organization authorized to reprocess the goods.

In January 2017, the *Ministère de l'Environnement, de l'Énergie et de la Mer* (French Ministry of the Environment, Energy and the Sea), in charge of international climate relations, issued a document on environmental taxation in France. In this respect, it should be noted that environmental taxes in France generated 47 billion euros in 2015. These €47 billion were divided into four categories: €38.5 billion for Energy (TICPE and related taxes), €6.1 billion for Transport (gray cards, insurance, civil aviation, etc.), €2.4 billion for Pollution (water and other pollution) and €0.4 billion for Resources (water sampling and others)[15].

7.5.2. *Environmental directives and decrees*

Directives are issued by the European Union. With the exception of constitutional provisions, they are at a superior level to the norms of French law. The decrees are drafted in France and are the responsibility of the executive branch represented by the President of the Republic and the government[16]. Decrees are the national transposition of directives.

Many texts have an impact on the logistics profession, the most famous of which are listed in the following.

14 http://www.douane.gouv.fr/articles/a12259-transporteurs-routiers-remboursement-des-taxes-de-carburant.

15 file:/thema_-_fiscalite_environnementale_-_un_etat_des_lieux.pdf.

16 http://infosdroits.fr/la-hierarchie-des-normes-dans-le-systeme-juridique-francais-principe-bloc-constitutionnel-legislatif-reglementaire-actes-conventionnel/.

7.5.2.1. *Directive 2000/53/EC on end-of-life vehicles*

EU Directive no. 2000/52/EC of September 18, 2000 applies to passenger vehicles and small trucks. Therefore, it does not (yet!) concern heavy goods vehicles. The Directive stresses that *"by January 1, 2015 at the latest, for all end-of-life vehicles, the rate of re-use and recovery shall be increased to a minimum of 95% by average weight per vehicle per year. Within the same period, the rate of reuse and recycling shall be increased to a minimum of 85% by average weight per vehicle per year"*[17].

7.5.2.2. *Directive 2002/95/EC on the management of waste electrical and electronic equipment*

Directive 2002/95/EC or RoHS, for Restriction of the use Of certain Hazardous Substances in electrical and electronic equipment, concerns the management of WEEE or waste electrical and electronic equipment. It is followed by Directives 2002/96/EC, 2011/65/EU and 2012/19/EU. Transposition in France appears with Decree no. 2005-829 of July 20, 2005. At present, Decree no. 2014-928 of August 19, 2014 is the authoritative decree. This decree specifies the obligations for manufacturers to take back electrical and electronic equipment and their reprocessing. Among the new elements of this regulation, collection targets are set so that *"from 2016 onwards, collection targets are set at 45% of the average weight of EEEs placed on the market over the last three years. In 2019, the target increases either to 65% of the average weight of EEEs placed on the market in the last three years, or to 85% of WEEE produced by weight"*[18].

For the treatment of WEEE, private eco-organizations which are authorized by the public authorities to carry out this reprocessing task are called upon. For WEEE, the eco-organizations are Ecologic (electrical and electronic equipment), PV Cycle (solar equipment), ESR (electrical, electronic and lamp equipment)[19] and ERP Recycling.

17 https://aida.ineris.fr/consultation_document/1003.

18 https://www.ecologique-solidaire.gouv.fr/dechets-dequipements-electriques-et-electroniques.

19 On June 30, 2017, the eco-organizations Eco-Systems (electrical and electronic devices) and Récylum (used lamps) were grouped together. They became ESR on January 1, 2018 (source: *L'Usine Nouvelle*, available at: http://www.usinenouvelle.com/article/les-eco-organismes-eco-systemes-et-recylum-fusionnent.N560173).

7.5.2.3. *Decree 2012-22 on waste furniture elements*

Decree 2012-22 of January 6, 2012 specifies the procedures for the *gestion des Déchets d'Éléments d'Ameublement (DEA)* (management of Furniture Element Waste). It is enriched by Decree no. 2016-288 of March 10, 2016 on the financing of DEAs through the Ecotax paid by the end consumer.

Three eco-organizations are authorized until December 31, 2017 for the prevention, collection and recovery of DEAs: EcoMobilier, Valdélia and Ecologic[20].

7.6. Sheet 41: Measuring Performance in Logistics

Key point

ASLOG points out that there is no standard for calculating logistics costs. Consequently, a comparison between companies is difficult. Also for ASLOG, the overall logistics costs are made up *of "transport costs on supply, raw material storage costs, financial costs of raw material stock, internal production logistics costs, financial costs of work-in-progresses and finished product stocks, downstream transport, storage costs of finished products, financial costs of finished product stocks linked to this storage"*[21]. However, performance measurement goes further, because it also takes non-financial indicators into account. Furthermore, it often appears that logistics dashboards are composed on average of 20% of financial indicators and 80% of non-financial indicators linked to logistics [MOR 13].

7.6.1. *The notion of quality and quantity indicators*

Two types of indicators are used in logistics: quality or non-financial indicators, and quantity or financial indicators. The quality indicators are measured by weight, meter, number of days and percentage. They are predominant in logistics because their link with quality, lead time and customer satisfaction is crucial. Quantity indicators have a monetary unit as

20 https://www.ecologique-solidaire.gouv.fr/dechets-delements-dameublement-dea.
21 http://www.web-logistique.com/nouvellepage1.htm.

the unit of value. They mainly come from accounting documents. This includes, for example, sales, results and leeway.

7.6.2. *Indicator themes*

There is no defined list of indicators in logistics. It is the company that lists the indicators it needs, in order to assess the performance of its structure. From an inter-organizational perspective, it sets out global indicators, which are generally used by many companies.

The indicators advocated in logistics are relative:

– costs. This includes transportation costs, storage costs, production costs and labor costs. More macro-economic indicators such as turnover or leeway can also be included. At this stage, the analysis is more likely to be by customer, customer segment, product or product line;

– quality. The search for a defect-free product is a prerequisite. The indicators around this notion are therefore many. At this level, it is also possible to insert indicators related to waste, or rather waste reduction, albeit more and more companies are including such indicators in an "environment" axis when they decide to establish a dashboard by axis (see next section);

– to the deadline. Delivering the product on time is a guaranteed way of keeping the customer. Here you can define an indicator on the delivery deadlines that were met. If necessary, this indicator format is specified by partial and/or complete delivery. In this way, partial deliveries are sometimes made to satisfy a customer and the remainder of the order is delivered at a later date. However, it should be noted that indicators on partial or complete deliveries are often included in the "quality" category;

– to the environment. More and more logistics indicators are referenced with pollution (particularly urban logistics indicators) and waste management [MOR 15];

– socially. Here, we find indicators such as absenteeism, accident, turnover and training rates.

7.6.3. *The dashboard*

The dashboard is a management tool capable of improving change within companies by introducing *"physical indicators, indicators that are not produced by the organization (unlike accounting and budgetary figures), environmental indicators, or even cross-cutting indicators"* [CHI 94]. It is a summary document that proposes the performance indicators deemed useful and relevant to an account of the company's situation. A dashboard can foresee an "indicator of excellence" to be obtained, in order to consider future actions of improvement. It may possibly refer to the best rates in the field of a catch-up strategy.

Title	Competition (a)	Desired (b)	Obtained (c)	Difference (c – a)	Difference (c – b)	Com-ments
Rate of service*	99% **	98%	97%	–2	–1	

Legend:
**The proposed indicator should be defined and described in an easily accessible tab if the document is computerized, or in an appendix if the document is in paper format.*
***Where the scoreboard refers to a rate of competition, it must first be verified that the method of calculating competition is identical to the method of calculation used in the undertaking.*

Table 7.1. *Example of a dashboard structure*

Two dashboard structures are currently under discussion in logistics: the structure of the SCOR® model and the balanced scorecard model.

7.6.3.1. *The SCOR® model structure*

Introduced in 1996, the SCOR® model (Supply Chain Operations Reference model®) is the product of the Supply Chain Council (SCC[22]), a grouping of Pittiglio, Rabin, Todd and McGarrah[23], AMR Research[24] and, initially, 69 companies from various economic sectors.

22 The Supply Chain Council was formed in 1996–97 on the initiative of PRTM, AMR Research, Procter and Gamble, Texas Instruments and other organizations. Its role is to promote and develop the SCOR® model (source: http://www.apics.org/apics-for-business/products-and-services/apics-scc-frameworks/scor).

23 PRTM (Pittiglio, Rabin, Todd and McGarrah) is a consulting company created in 1976, with the aim of providing technological assistance to various international companies (source: www.prtm.com).

The metrics structure of the SCOR® model is based on four levels, ranging from a strategic level to an operational level (see Sheet 3, section I.3.3). Level 1 metrics are available on the APICS website, the only body authorized to disseminate the model. On this level, five areas are listed, namely (1) responsiveness, with indicators linked to the "order execution cycle", (2) agility, with indicators that address the benefits of supply chain flexibility and adaptability, declining supply chain adaptability and total value at risk, (3) cost, with cost indicators on each process in the supply chain, and (4) efficiency of asset management with indicators on the duration of the operating fund cycle, return on investment and return on working capital[25].

7.6.3.2. *The balanced scorecard model*

From across the Atlantic, the balanced scorecard was initiated by Kaplan and Norton [KAP 92, KAP 96]. Its architecture is presented in a multidimensional aspect, through the implementation of axes, where a certain number of indicators are then declined. Originally, these two authors proposed four axes of measurement: (1) financial, (2) internal processes, (3) clients and (4) organizational learning and innovation.

This division was criticized, either because it lacked axes, such as a supplier axis, or because some axes were not necessary on an operational level, such as the innovation axis.

Today, the division by axis is still relevant, even if some people modify it to adapt it to the company's needs. For example, Morana and Gonzalez-Feliu [GON 15] proposed an urban logistics dashboard with three axes based on Sustainable Development, with an economic axis, an environmental axis and a social axis (see Table 7.2)

24 AMR (Advanced Manufacturing Research), founded in 1986, is a consulting company specialized in the programming of process management measurement systems (source: www.amresearch.com).

25 http://www.apics.org/docs/default-source/scc-non-research/apicsscc_scor_quick_reference_guide.pdf.

Axis	Category	Key Indicators (total = 8)
Economical	Logistics	Operating costs
	Audit	Monetary savings
Economical	Quality service	Rate of service
		Customer satisfaction rate
		Customer development
Environmental	Environmental effects	Pollution emission rate
Environmental	Reduction of congestion	Congestion rate
Social/societal	Social/societal effects	Number of jobs created, destroyed and reconverted

Table 7.2. *Urban logistics dashboard (source: Morana and Gonzalez-Feliu [MOR 15])*

Conclusion

C.1. Sheet 42: Conclusion – Some Atypical Logistics

Key point

Logistics is a "rich" profession because it is constantly evolving and expanding. The logistician must possess diverse skills, not only in his own field, but also in areas that support logistics. We conclude this overview of logistics by presenting some "atypical" logistics.

C.1.1. *Hospital logistics*

Hospital logistics is defined as *"a set of design, planning and delivery activities that enable the procurement, inventory management and replenishment of goods and services surrounding the benefit of medical services to patients"*. [LAN 00, LAN 02]. It is also perceived as *"an open whole crossed by human, material, informational and financial flows, made up of various autonomous entities (suppliers, hospital services (emergency, operating room, pharmacy, cardiology...), logistics providers, medical providers...) that use limited resources (time, material, assets, workers) and coordinate their actions through an integrated logistics process in order to improve their collective performance first and foremost (patient satisfaction, global optimization of hospital operations), but also their individual performance in the long term (maximization of an entity's profit)"* [FÉN 04, FÉN 05, AND 07].

Hospital logistics is considered important for strategic decisions of hospital management because of:

– Increasing financial pressure on healthcare institutions. For control bodies, such as the *Agences Régionales de Santé en France* (Regional Health Agencies in France), hospitals must make efforts to reduce healthcare production costs. In these cases, logistics is a significant source of cost savings.

– Changes in regulations and certifications, which oblige hospitals to adapt to new rules: certification procedures, nominal dispensation of medicines, tracking of medicines, etc.

– Technological evolutions that lead healthcare institutions to modify their logistics tools: barcodes to ensure tracking, development of IT tools to manage stock and supplies, etc.

Two main themes are dealt with in hospital logistics, namely:

– the issue of scheduling and planning in hospitals: patient expectations [CHA 11]; management of services, particularly the operating theatre [CHA 07];

– the optimization of stocks of medicines and other products (sheets, food, maintenance, etc.) [MON 11, MOU 13].

C.1.2. *Humanitarian logistics*

Humanitarian logistics is associated with a context of crisis: natural disaster, geopolitical, famine and health problems. It requires a two-step action. First (and very quickly), emergency actions to deal with the survival of people who are facing disasters, and then, in the longer term, reconstruction actions. As Chandes and Paché [CHA 06] point out, humanitarian logistics is – unlike other types of logistics – the subject of little scientific study. However, even if it can find inspiration in hospital logistics work (given that one of the main elements of a humanitarian crisis is the "construction" of an emergency hospital), its specifications require special attention.

According to Thomas [THO 03], Trunick [TRU 05] and Giroux *et al.* [GIR 09], 80% of humanitarian actions are logistical in nature. Giroux *et al.* [GIR 09] classify humanitarian logistics as "extreme logistics". In addition, according to Lièvre and Gautier [LIÈ 09], an extreme situation presents *"a certain split from its usual, everyday life [RIV 98]. [....] The context appears to be uncertain [...] and we are dealing with a risky situation, that is to say a situation in which the possibility of an unwanted event occurs and causes more or less significant damage to the organization"*. From this perspective, humanitarian logistics is part of the logistics of extreme situations.

With regard to the country and the place of intervention, there may be a greater or lesser number of people involved in humanitarian action – and by extension in humanitarian logistics – with international impacts. Chandes and Paché [CHA 06] summarize four types of speakers:

– *"NGOs of different nationalities, sizes and skill levels (medical, logistical, nutritionists, emergency aid and/or development specialists, etc.);*

– bilateral operators: embassies, aid workers, civil protection, army, fire brigades, etc;

– multilateral operators such as ECHO[1]*;*

– international organizations of the United Nations or the International Committee of the Red Cross".

C.1.3. *Event logistics*

An event can be assessed by four criteria: the public concerned (public, operators, technicians, etc.), the venue (a dedicated space), a generally limited timeframe and a rarity [VAU 09]. According to Salaun [SAL 16], event logistics can be assessed in terms of (1) *"the compression of action time"* and (2) *"the diversity of flows to be managed"*.

Highly insightful specialists

The term "event" covers several things, among the main ones are: thematic shows, fairs, exhibitions, but also conventions and sometimes conferences. Event logistics consists of moving all the necessary elements to the venue where the event will take place and then, once it is over, returning them to their original location or, in the case of a tour, transporting all the equipment from town to town. This is very similar to transporting a show tour and event logistics providers often provide this type of service. Specialists experienced in this practice generally do not only place

1 ECHO: European Commission Humanitarian Office – http://ec.europa.eu/echo/index_fr

the appropriate materials in an unloading area, but also deliver each item to the specific location where it will be displayed and, if necessary, handle it.

Event logistics worldwide

With the multiplication of exhibitions and conventions on international tours, a worldwide event logistics service has developed. They are often former companies that were specialized in the transport of sets for musicals and operas. Ultimately, only the scale varies with regard to this activity and sometimes the modes of transport, particularly maritime transport. These bodies then take care of the formalities for customs clearance between countries and the storage of the items, before they are dispatched or re-dispatched.

Box C.1. *What is event logistics? (source: http://www.communique-2-presse.com/quest-logistique-evenementielle/ – 13/11/2015)*

In 2015, the economic benefits of fairs, exhibitions and conventions in France accounted for 7.5 billion euros for 120,000 jobs [UNI 17]. The implementation of an event can be associated with obtaining labels such as the ISO 20121:2012 standard, which provides guidelines for managing an event under a sustainable development approach: *"Event activities take a heavy toll on our resources, society in general and the environment. They can produce large amounts of waste, overuse local resources such as water or energy sources, or even create tensions within communities. Thanks to ISO 20121, your event, whatever its type or size, can be part of a sustainable development logic"*[2].

2 https://www.iso.org/fr/iso-20121-sustainable-events.html

Bibliography

I.1. Sheet 1: The Logistics Function

[BAZ 08] BAZIN S., BECKERICH C., "Le renforcement des contraintes spatiales des entreprises: le transport comme vecteur de valorisation de la prestation logistique", *Les Cahiers Scientifiques du Transport*, no. 54, pp. 49–72, 2008.

[BER 08] BERNADET M., "Transport et logistique durables: vers la nécessaire prise en compte de leur interdépendance", *Notes de synthèse du SESP*, no. 168, pp. 5–8, 2008.

[BLA 08] BLANQUART C., CARBONE V., "La place du transport dans les organisations logistiques durables", *Les Cahiers Scientifiques du Transport*, no. 54, pp. 11–24, 2008.

[BLA 10] BLANQUART C., CARBONE V., "Pratiques collaboratives et démarche environnementale dans la supply chain : mythe ou réalité?", *8èmes Rencontres Internationales de la Recherche en Logistique*, Bordeaux, France, 2010.

[CAR 08] CARTER C., ROGERS D., "A framework of sustainable supply chain management: moving toward new theory", *International Journal of Physical Distribution & Logistics Management*, vol. 38, no. 5, pp. 360–387, 2008.

[CHR 92] CHRISTOPHER M., *Logistics and Supply Chain Management*, Pitman Publishing, London, 1992.

[COL 03] COLLIN M. (ed.), *Ville portuaire, acteur du développement durable*, L'Harmattan, Paris, 2003.

[DUR 10] DURAND B., Quand la durabilité interpelle l'e-logistique…., Working Paper, no. EA 4272, Lemna, Nantes, 2010.

[FAB 94] FABBE-COSTES N., COLIN J., "Formulating logistics strategy", in COOPER J. (ed.), *Logistics and Distribution Planning*, Kogan Page, London, 1994.

[GIA 03] GIARD V., *Gestion de la production et des flux*, Economica, Paris, 2003.

[GON 10] GONZALEZ-FELIU J., MORANA J., "A la recherche d'une mutualisation des livraisons en milieu urbain: le cas du groupe NMPP", *Revue Française de Gestion Industrielle*, vol. 29, no. 2, pp. 71–92, 2010.

[LAM 98] LAMBERT D., COOPER M., PAGH J., "Supply chain management: implementation issues and research opportunities", *The International Journal of Logistics Management*, vol. 9, no. 2, pp. 1–19, 1998.

[MAS 10] MASSON S., PETIOT R., "Attractivité territoriale, infrastructures logistiques et développement durable", *Colloque ASRDLT – AIRSe*, Aoste, 20–22 September, 2010.

[MEN 01] MENTZER J., DEWITT W., KEEBLER J. *et al.*, "Defining supply chain management", *Journal of Business Logistics*, vol. 22, no. 2, pp. 1–25, 2001.

[MOR 13] MORANA J., *La logistique durable*, Hermes Science-Lavoisier, Paris, 2013.

[PAG 09] PAGELL M., WU Z., "Building a more complete theory of sustainable supply chain management using case studies of 10 exemplars", *Journal of Supply Chain Management*, vol. 45, no. 2, pp. 37–56, 2009.

[PAN 10] PAN S., Contribution à la définition et à l'évaluation de la mutualisation de chaînes logistiques pour réduire les émissions de CO2 du transport : application au cas de la grande distribution, PhD thesis, Ecole nationale supérieure des mines de Paris, 2010.

[SAV 10] SAVY M., LIU X., "La plate-forme logistique, objet exemplaire d'aménagement urbain", *Les premières Journées du Pôle Ville de l'Université Paris-Est Ville: Transport et Territoire, Quoi de neuf?*, PRES Paris-Est, 2010.

[SEU 08] SEURING S., MÜLLER M., "From a literature review to a conceptual framework for sustainable supply chain management", *Journal of Cleaner Production*, vol. 16, no. 14, pp. 1699–1710, 2008.

[SVE 07] SVENSSON G., "Aspects of sustainable supply chain management (SSCM): conceptual framework and empirical example", *Supply Chain Management: An International Journal*, vol. 12, no. 4, pp. 262–266, 2007.

I.2. Sheet 2: Supply Chain Flows

[ARN 98] ARNOLD T., *Introduction to materials management*, Prentice-Hall, Englewood Cliffs, 1998.

[AUR 97] AURIFEILLE J.-M., COLIN J., FABBE-COSTES N. *et al.*, *Management logistique: une approche transversale*, Litec, Paris, 1997.

[BIG 95] BIGRAS Y., HALLEY A., GÉLINAS R., "Juste-à-temps et transport dans le marché nord-américain", *Actes de la 30e Conférence annuelle du Canadian Transportation Research Forum*, Aylmer, Canada, 15–17 May 1995.

[COL 88] COLIN J., PACHÉ G., *La logistique de distribution: l'avenir du marketing*, Chotard et Associés Editeurs, Paris, 1988.

[FAB 97] FABBE-COSTES N., "L'intervention de la logistique dans la formulation/mise en acte de la stratégie en milieu complexe", in AVENIER M.-J. *et al.* (eds), *La stratégie "chemin faisant"*, Economica, Paris, 1997.

[GAR 08] GARCIA-DASTUGUE S.J., LAMBERT D.M., "Différenciation retardée interorganisationnelle dans la supply chain", *Logistique & Management*, vol. 16, no. 1, pp. 31–46, 2008.

[HES 73a] HESKETT J., IVIE R., GLASKOWSKY N., *Business Logistics*, The Ronald Press, New York, 1973.

[HES 73b] HESKETT J., "Sweeping changes in distribution", *Harvard Business Review*, vol. 51, no. 2, pp. 123–133, 1973.

[LAM 04] LAMBERT S., RIOPEL D., "Modèle intégrateur de la logistique inverse", *Les Cahiers du GERAD*, no. G-2004-69, August 2004.

[MES 99] MESNARD X., DUPONT A., "Votre logistique est-elle à la pointe?", *L'Expansion Management Review*, no. 94, pp. 52–58, 1999.

[PAC 99] PACHÉ G., SAUVAGE T., *La logistique: enjeux stratégiques*, Vuibert Entreprise, Paris, 1999.

[PET 06] PETITQUEUX A., "Implementation Lean: application industrielle", *Les Techniques de l'Ingénieur*, AG5195v1, 2006.

[TIX 79] TIXIER D., "Les perspectives de la logistique d'entreprise", *Revue Française de Gestion*, no. 19, pp. 85–89, 1979.

[TIX 83] TIXIER D., MATHE H., COLIN J., *La logistique au service de l'entreprise*, Dunod, Paris, 1983.

I.3. Sheet 3: The Main Models of the Global Supply Chain

[BOW 99] BOWERSOX D., CLOSS D., STANK T., "21st century logistics: making supply chain integration a reality", *Council of Logistics Management*, Oak Brook, 1999.

[CHR 92] CHRISTOPHER M., *Logistics and Supply Chain Management*, Pitman Publishing, London, 1992.

[CHR 94] CHRISTOPHER M., "New direction in logistics", in COOPER J. (ed.), *Logistics and Distribution Planning*, Kogan Page, London, 1994.

[CHR 97] CHRISTOPHER M., *Marketing Logistics*, Butterworth-Heinemann, Oxford, 1997.

[CHR 99] CHRISTOPHER M., "Les enjeux d'une supply chain globale", *Logistique & Management*, vol. 7, no. 1, pp. 3–6, 1999.

[CHR 00] CHRISTOPHER M., "The agile supply chain: competing in a volatile markets", *Industrial Marketing Management*, vol. 29, no. 1, pp. 37–44, 2000.

[GLR 95] GLRT – GLOBAL LOGISTICS RESEARCH TEAM, "World Class Logistics: the challenge of managing continuous change", *Council of Logistics Management*, Oak Brook, 1995.

[LAM 98] LAMBERT D., COOPER M., PAGH J., "Supply chain management: implementation issues and research opportunities", *The International Journal of Logistics Management*, vol. 9, no. 2, pp. 1–19, 1998.

[MEN 01] MENTZER J., DEWITT W., KEEBLER J. *et al.*, "Defining supply chain management", *Journal of Business Logistics*, vol. 22, no. 2, pp. 1–25, 2001.

[MIN 04] MIN S., MENTZER J., "Developing and measuring supply chain concepts", *Journal of Business Logistics*, vol. 25, no. 1, pp. 63–99, 2004.

[SCO 96] SCOR FRAMEWORK, http://www.apics.org/sites/apics-supply-chain-council/frameworks/scor, 1996.

1.1. Sheet 5: Transport and Logistics Jobs

[PAG 15] PAGE M., ASLOG, Etude de Fonctions et Rémunérations des Métiers de la Logistique et de la Supply Chain, available at: http://etudes-presse.michaelpage.fr/etudes/etude-2015-de-fonctions-et-remunerations-des-metiers-de-la-logistique-et-de-la-supply-chain-2/, 2015.

1.2. Sheet 6: The Supply Chain Manager

[LIV 11] LIVOLSI L., "Profiling the job of supply chain manager: a French case study", *Supply Chain Forum: an International Journal*, vol. 12, no. 3, pp. 54–63, available at: http://www.supplychain-forum.com/documents/articles/SCFIJVol 12-3-2011-Livolsi.pdf, 2011.

[POR 86] PORTER M., *L'avantage concurrentiel*, Dunod, Paris, 1986.

1.3. Sheet 7: Logistics and Interactions with Other Internal Departments of the Company

[ELK 10] EL KORCHI A., MILLET D., "Conditions stratégiques d'émergence d'une reverse supply chain fondée sur le remanufacturing chez le fabricant d'origine", *8èmes Rencontres Internationales de Recherche en Logistique et SCM*, Bordeaux, France, 29–30 September 2010.

[MIC 10] MICHAUD C., Comportements des consommateurs et biens privés à caractéristiques environnementales: une approche expérimentale, PhD thesis, University of Grenoble, 1 March 2010.

[WAG 89] WAGONER S., "Logistics and quality management", *Logistics Spectrum*, vol. 23, no. 4, pp. 13–16, 1989.

1.4. Sheet 8: The Supplier-Customer Relationship

[GAR 03] GARDNER J., COOPER M., "Cartographie stratégique de la chaîne logistique", *Logistique & Management*, vol. 11, no. 2, pp. 55–73, 2003.

[LAM 98] LAMBERT D., COOPER M., PAGH J., "Supply chain management: Implementation issues and research opportunities", *The International Journal of Logistics Management*, vol. 9, no. 2, pp. 1–19, 1998.

[MIN 02] MIN H., ZHOU G., "Supply chain modeling, past, present and future", *Computers & Industrial Engineering*, vol. 43, nos 1–2, pp. 231–249, 2002.

[PAC 00] PACHÉ G., COLIN J., "Recherche et applications en logistique: des questions d'hier, d'aujourd'hui et de demain", in FABBE-COSTES N., CCLIN J., PACHÉ G. (eds), *Faire de la recherche en logistique et distribution?*, Vuibert-Fnege, Paris, 2000.

[PHI 97] PHILIPPE J., "Relations producteur-distributeur: les enseignements du secteur de l'habillement", *Economie et statistique*, no. 303, pp. 101–112, March 1997.

1.5. Sheet 9: Operators of Distribution

[CAP 13] CAPON G., FUNEL C., SURY R., "Les grossistes, acteurs majeurs des mutations de la distribution industrielle – Analyse de la rentabilité et des modèles économiques du négoce des produits industriels", *Cahier de recherche du Crédoc*, no. 306, December 2013.

[CHA 12] CHANUT O., PACHÉ G., WAGENHAUSEN F., "Logistique urbaine: refonder les logiques d'intermédiation", *Management & Avenir*, no. 51, pp. 186–207, 2012.

[DUG 00] DUGOT PH., *Le commerce de gros*, Presses Universitaires de France, Paris, 2000.

[DUR 10] DURAND B., GONZALES-FELIU J., HENRIOT F., "La logistique urbaine, facteur-clé de développement du B to C", *8ème Rencontres Internationales de la Recherche en Logistique*, Bordeaux, France, 29–30 September and 1 October 2010.

[FIL 01] FILSER M., DES GARETS V., PACHÉ G., *La distribution: organisation et stratégie*, EMS Management & Société, Cormelles-Le-Royal, 2001.

[PAC 08] PACHÉ G., "Efficient urban e-logistics: mutualization of resources and source of competitive advantage", *7th International Meeting for Research in Logistics*, Avignon, France, 24–26 September 2008.

[PHI 97] PHILIPPE J., "Relations producteur-distributeur : les enseignements du secteur de l'habillement", *Economie et statistique*, no. 303, pp. 101–112, March 1997.

1.6. Sheet 10: Logistics Service Providers

[FUL 11] FULCONIS F., PACHÉ G., ROVEILLO G., *La prestation logistique – origines, enjeux et perspectives*, EMS Management & Société, Cormelles-Le-Royal, 2011.

[HAL 04] HALLEY A., "Sous-traitance et chaîne logistique ou la nécessaire intégration des deux strategies", *Gestion*, vol. 29, no. 2, pp. 48–56, 2004.

[LEX 04] LEXIPRO, Le lexique des termes de la logistique, *Logistiques magazine*, October 2004.

[ROQ 03] ROQUES T., MICHRAFY, "Logistics service providers in France – 2002 survey: actors' perceptions and changes in practice", *Supply Chain Forum: an International Journal*, vol. 4, no. 2, pp. 34–52, 2003.

[VON 13] VON PFUHLSTEIN H., Kuehhe + Nagel: Challenges and adaptation of the operating model, Powerpoint presentation, p. 19, available at: http://www.rolandberger.ch/media/pdf/Roland_Berger_OperatingModelKN20131107.pdf, November 2013.

1.7. Sheet 11: Operators in the Transport of Goods for Third Parties

[ADE 17] ADEME, Stratégie, Transports et Mobilité – Période 2014-2017, available at: https://www.ademe.fr/sites/default/files/assets/documents/strategie-transports-et-mobilite-periode-2014-2017.pdf, 2017.

[BER 06] BERNABET M., "Les professions du transport", *Techniques de l'Ingénieur,* document AG 8 020, 2006.

[LEG 08] LEGRAND G., MARTINI H., *Gestion des opérations Import-Export*, Dunod, Paris, 2008.

[OBE 87] OBSERVATOIRE ECONOMIQUE ET STATISTIQUE DES TRANSPORTS, Les auxiliaires de transport : points de repère, p. 4, available at: http://temis.documentation.developpement-durable.gouv.fr/documents/temis/NS/NS_008_2.pdf, 1987.

1.8. Sheet 12: Operators in Urban Logistics

[CER 14] CEREMA, *La logistique urbaine-connaître et agir*, Territoires et Villes, Collection Références, France, 2014.

[CHA 12] CHANUT O., PACHÉ G., WAGENHAUSEN F., "Logistique urbaine: refonder les logiques d'intermédiation", *Management & Avenir*, no. 51, pp. 186–207, 2012.

[DUC 14] DUCRET R., Nouvelles organisations de la distribution urbaine des colis sur le dernier kilomètre : innover par une approche spatiale, PhD thesis, Ecole Nationale Supérieure des Mines de Paris, 19 December 2014.

1.9. Sheet 13: Reverse Logistics Operators

[CAR 13] CARNEIRO J., MONNET M., NOIREAUX V., "Quelle maturation de la logistique inversée?", in FABBE-COSTES N., PACHÉ G. (eds), *La logistique – Une approche innovante des organisations*, Presses Universitaires de Provence, Aix-en-Provence, 2013.

[FLE 97] FLEISCHMANN M., BLOEMHOF-RUWAARD J., DEKKER R. *et al.*, "Quantitative model for reverse logistics: a review", *European Journal of Operational Research*, vol. 103, no. 1, pp. 1–17, 1997.

[FRE 16] FRENCH MINISTRY OF THE ENVIRONMENT, Energy and the Sea, Filières de responsabilité élargie du producteur (REP), Prévention des risques, available at: http://www.developpement-durable.gouv.fr/Dechets-d-elements-d-ameublement.html, 26 January 2016.

[LAM 03] LAMBERT S., RIOPEL D., "Logistique inverse: revue de littérature", *Les Cahiers du GERAD*, no. G-2003-61, October 2003.

[MON 11] MONNET M., "Les stratégies de logistique inversée: une perspective théorique", *Logistique & Management*, vol. 19, no. 1, pp. 41–54, 2011.

2.1. Sheet 14: Different Types of Innovation

[DAN 05] DANDURAND L., "Réflexion autour du concept d'innovation sociale, approche historique et comparative", *Revue Française d'Administration Publique*, vol. 3, no. 115, pp. 377–382, 2005.

[OSL 05a] OSLO MANUAL, La mesure des activités scientifiques et technologiques, 3rd edition, available at: http://www.oecd.org/dataoecd/35/58/2367554.pdf, 2005.

[OSL 05b] OSLO MANUAL, Principes directeurs pour le recueil et l'interprétation des données sur l'innovation, 3rd edition, available at: http://www.oecd.org/fr/science/inno/manueldosloprincipesdirecteurspourlerecueiletlinterpretationdesdonneessurlinnovation3eedition.htm, 2005.

2.2. Sheet 15: The Concept of Eco-innovation and Its Application in Logistics

[CHA 11] CHARDINE-BAUMANN E., Modèles d'évaluation des performances économique, environnementale et sociale dans les chaînes logistiques, PhD thesis, Institut National des Sciences Appliquées de Lyon, May 2011.

[FER 13] FERNANDES V., VAN STEENBERGER M., "Le management durable de la supply chain : quelles pratiques pour réduire l'impact environnemental d'un site industriel?", *Gestion 2000*, vol. 30, no. 1, pp. 53–67, 2013.

[FOU 12] FOULQUIER L., "Le parcours des mots: le cas du préfixe "éco" et l'écologie", *Environnement, Risques et Santé*, vol. 11, no. 3, pp 230–239, May–June 2012.

[KIN 13] KING K., "Strategic innovation priorities for sustainable manufacturing in Australia", *11th Conference on Sustainable Manufacturing*, Berlin, 2013.

[MAT 10] MATHIEU A., CHANDON J.-L., REYNAUD E., "Le développement durable en action: approche par l'éco-innovation", *Actes de la XIXème Conférence de l'Association Internationale de Management Stratégique*, Luxembourg, 2–4 June 2010.

[NEW 96] NEWMAN W., HANNA M., "An empirical exploration of the relationship between manufacturing strategy and environmental management: two complementary models", *International Journal of Operations & Production Management*, vol. 16, no. 4, pp. 69–87, 1996.

[OEC 09] OECD, Sustainable Manufacturing and Eco-Innovation: Framework, Practices and Measurement 2009, available at: https://www.oecd.org/innovation/inno/43423689.pdf, 2009.

[ROG 01] ROGERS D., TIBBEN-LEMBKE R., "An examination of reverse logistics practices", *Journal of Business Logistics*, vol. 22, no. 2, pp. 129–149, 2001.

2.3. Sheet 16: Sustainable Logistics Innovation

[ACO 13] ACOSTA P., ACQUIER A., CARBONE V. *et al.*, Innovation + développement durable = nouveaux business models, Etude Innovation Durable – Institut i7 – ESCP Europe – weave AIR, p. 88, available at: http://institut-innovation-competitivite.eu/sites/default/files/innovation_durable_i7_escp_europe_weave_air_2013_0.pdf, June 2013.

[ADE 13] ADEME, Economie circulaire : notions, Direction Economie Circulaire et Déchets – ADEME Angers, p. 10, available at: http://www.ademe.fr/sites/default/files/assets/documents/fiche-technique-economie-circulaire-oct-2014.pdf, 2013.

[VAN 14] VAN NIEL J., "L'économie de la fonctionnalité: principes, éléments de terminologie et proposition de typologie", *Développement Durable et Territoires*, vol. 5, no. 1, available at: https://developpementdurable.revues.org/10160, February 2014.

3.1. Sheet 17: Design of a Warehouse

[AUM 08] AUMAS M., Les rayonnages métalliques, ED 771, Institut National de Recherche et de Sécurité (INRS), 2008.

[BAK 09] BAKER P., CANESSA M., "Warehouse design: a structured approach", *European Journal of Operational Research*, vol. 193, no. 2, pp. 425–436, 2009.

[DEK 07] DEKOSTER R., LE-DUC T., JAN ROODBERGEN K., "Design and control of warehouse order picking: A literature review", *European Journal of Operational Research*, vol. 182, no. 2, pp. 481–501, 2007.

[GOE 89] GOETSCHALCKX M., ASHAYERI J., "Classification and design of order picking systems", *Logistics World*, vol. 2, no. 2, pp. 99–106, June 1989.

[GU 07] GU J., GOETSCHALCKX M., MCGINNIS L.F., "Research on warehouse operation: A comprehensive review", *European Journal of Operational Research*, vol. 177, no. 1, pp. 1–21, 2007.

[GU 10] GU J., GOETSCHALCKX M., MCGINNIS L.F., "Research on warehouse design and performance evaluation: a comprehensive review", *European Journal of Operational Research*, vol. 203, no. 3, pp. 539–549, 2010.

[LAU 10] LAUDON A.J., RIOPEL D., Analyse préparatoire de données pour un projet de conception d'entrepôt centralisé du système de santé québécois, CIRRLET-2010-41, 2010.

[ROG 09] ROGNON J.-L., "Stockage: ne prenez pas vos palettiers à la légère", *Supply Chain Magazine,* no. 40, pp. 68–76, available at: http://www.supply chainmagazine.fr/TOUTE-INFO/Archives/SCM040/Appels-offre-40.pdf, 2009.

[ROO 09] ROODBERGEN K.J., VIS I.F.A., "A survey of literature on automated storage and retrieval systems", *European Journal of Operational Research*, vol. 194, no. 2, pp. 343–362, 2009.

[ROU 00] ROUWENHORST B., REUTER B., STOCKRAHM V. *et al.*, "Warehouse design and control: framework and literature review", *European Journal of Operational Research*, vol. 122, no. 3, pp. 515–533, 2000.

[ROU 06] ROUX M., "Réflexions sur les entrepôts et les magasins", *Revue Française de Gestion Industrielle*, vol. 25, no. 2, pp. 59–69, 2006.

[TRE 10] TREMBLAY E., BOCTOR F., RENAUD J., "Impacts de la configuration de l'entrepôt et de la localisation des produits sur l'efficacité des parcours de collecte: un cas reel", *ASAC 2010*, Université Laval, 2010.

3.2. Sheet 18: Layout of a Warehouse

[ROU 08] ROUX M., *Entrepôts et magasins*, Eyrolles, Paris, 2008.

3.3. Sheet 19: Types of stock

[DUR 07] DURAND B., "La loi Dutreil II vue sous l'angle logistique: l'émergence de nouveaux scénarios est-elle vraiment pour demain?", *1ère journée de recherche relations entre industrie et grande distribution*, Avignon, France, 29 March 2007.

3.5. Sheet 21: The Supply Quantity

[ERL 90] ERLENKOTTER D., "Ford Whitman Harris and the economic order quantity model", *Operations Research,* vol. 38, no. 6, pp. 937–946, available at: http://pubsonline.informs.org/doi/pdf/10.1287/opre.38.6.937, 1990.

[HAR 13a] HARRIS F.W., "How many parts to make at once", *Factory, The Magazine of Management*, vol. 10, pp. 135–136, 1913.

[HAR 13b] HARRIS F.W., "How much stock to keep on hand", *Factory, The Magazine of Management*, vol. 10, pp. 240–241 and pp. 281–284, 2013.

[HAR 15] HARRIS F.W., "What quantity to make at once", *The Library of Factory Management, vol. V: Operation and Costs*, A. W. Shaw Company, Chicago, pp. 47–52, 1915.

[WIL 27] WILSON R.H., MUELLER W.A., "A new method of stock control", *Harvard Business Review*, vol. 5, pp. 197–205, 1927.

[WIL 34] WILSON R.H., "Scientific routine for stock control", *Harvard Business Review*, vol. 13, pp. 116–128, 1934.

4.1. Sheet 23: The Relationship Between the Purchasing and Logistics Department

[BRU 08] BRUEL O., *Politique d'achat et gestion des approvisionnements*, Dunod, Paris, 2008.

[CAL 06] CALVI R., MERMINOD N., PACHÉ G., "Pour une approche décloisonnée de la logistique et des achats: proposition d'un modèle CCO", *Revue Française de Gestion Industrielle*, vol. 25, no. 3, pp. 29–46, 2006.

[CHA 11] CHARDINE-BAUMANN E., Modèles d'évaluation des performances économique, environnementale et sociale dans les chaînes logistiques, PhD thesis, Institut National des Sciences Appliquées de Lyon, May 2011.

[DUR 07] DURAND B., DE FAULTRIER B., "L'impact de la supply chain sur les métiers de la logistique et des achats", *Logistique & Management*, vol. 15, no. 2, pp. 55–70, 2007.

[FOR 58] FORRESTER J.W., "Industrial dynamics: a major breakthrough for decision makers", *Harvard Business Review*, vol. 36, no. 4, pp. 67–66, 1958.

[ROU 16] ROUQUET A., ROUSSAT CH., VEGA D., "Jay Wright Forrester: une première esquisse du Supply Chain Management", in LAVASTRE O., CARBONE V., AGERON B. (eds), *Les grands auteurs en logistique et supply chain management*, EMS – Management & Société, Cormelles-Le-Royal, pp. 85–98, 2016.

4.2. Sheet 24: Tendering Process, Specification and Logistics Service Contract

[CAR 04] CARBONE V., Le rôle des prestataires logistiques en Europe – intégration des chaînes et alliances logistiques, PhD thesis, Ecole Nationale des Ponts et Chaussées, France, 2004.

[DES 14] DES MESNARDS, "Cahier des charges fonctionnel", *Techniques de l'Ingénieur*, 0574, 25 March 2014.

[ROQ 03] ROQUES T., "L'achat de la prestation logistique et la mise en place d'un partenariat industriel-prestataire", *Logistique & Management*, vol. 11, no. 1, pp. 67–76, 2003.

5.1. Sheet 26: Key Concepts of Production Logistics

[LAM 99] LAMOURI S., THOMAS A., "Juste à temps et qualité totale : concepts et outils", *Techniques de L'ingénieur*, AG 5 190, 1999.

[LYO 15] LYONNET B., MARIE-PASCALE S., *La logistique*, Dunod, Paris, 2015.

5.2. Sheet 27: Lean Manufacturing Fundamentals

[CHI 14] CHICK S., HUCHZERMEIER A., NETESSINE S., "Usines européennes et excellence industrielle", *Harvard Business Review*, pp. 1–6, August–September 2014.

[IGA 01] IGALENS J., JORAS M., *La responsabilité sociale de l'entreprise*, Editions d'Organisation, Paris, 2001.

[LOP 15] LOPEZ F., Gemba, Workplace, Genchi Gembutsu, Go-and-see… What's the Difference?, available at: http://www.lean.org/balle/DisplayObject.cfm?o=3058, 2015.

[LOR 14] LORINO PH., La fuite managériale devant la complexité : l'exemple historique du "*lean management*", ESSEC Working Paper 1410, 2014.

[PET 06] PETITQUEUX A., "Implementation Lean: application industrielle", *Les Techniques de l'Ingénieur*, 2006.

[RÉG 03] RÉGOL O., BÉLANGER P.R., "Le KAIZEN : ses principes et ses conséquences pour les ouvriers et les syndicats. Volet 1: Revue de la littérature", *Les Cahiers du CRISES – Collection Etudes Théoriques*, ET0306, 2003.

[SUA 11] SUAREZ-BARRAZA M., RAMIS-PUJOL J., KERBACHE L., "Thoughts on Kaizen and its evolution: three different perspectives and guiding principles", *International Journal of Lean Six Sigma*, vol. 2, no. 4, pp. 288–308, 2011.

5.3. Sheet 28: Lean Manufacturing Tools

[AGA 04] AGARD B., "Conception et fabrication des familles de produits – Etat de l'art", *RE-JESA*, vol. 38, nos 1–2, pp. 59–84, 2004.

[DUD 09] DUDEK-BURLIKOWSKA M., SZEWIECZEK D., "The Poka-Yoke method as an improving quality tool of operations in the process", *Journal of Achievements in Materials and Manufacturing Engineering*, vol. 36, no. 1, pp. 95–102, 2009.

[FUL 05] FULCONIS F., PACHÉ G., "Piloter des entreprises virtuelles : Quel rôle pour les prestataires de services logistiques", *Revue Française de Gestion*, vol. 3, no. 156, pp. 167–186, 2005.

[GIA 07] GIARD V., MENDY G., "De l'approvisionnement synchrone à la production synchrone dans la chaîne logistique", *Revue Française de Gestion*, vol. 2, no. 171, pp. 65–88, 2007.

[KUS 87] KUSIAK A., "The generalized group technology concept", *International Journal of Production Research*, vol. 25, no. 4, pp. 561–569, 1987.

[LAS 07] LASNIER G., "Le *lean-manufacturing*", *La Revue des Sciences de Gestion*, vol. 1, no. 223, pp. 99–107, 2007.

[LYO 15] LYONNET B., SENKEL M.-P., *La logistique*, Dunod, Paris, 2015.

[PET 06] PETITQUEUX A., "Implementation Lean: application industrielle", *Les Techniques de l'Ingénieur*, 2006.

[SER 08] SERRANO LASA I., OCHOA LABURU C., DE CASTRO VILA R., "An evaluation of the value stream mapping tool", *Business Process Management Journal*, vol. 14, no. 1, p. 39, 2008.

[TAR 98] TARONDEAU J.-C., *Stratégie Industrielle*, Vuibert, Paris, 1998.

5.4. Sheet 29: Implementation of Workshops

[HAC 09] HACHICHA W., Nouvelles approches pour la formation des cellules de production dans le cadre d'une démarche de conception, PhD thesis, École nationale de Sfax, 13 March 2009.

[KIN 80] KING J.R., "Machine-component grouping in production flow analysis: an approach using a rank order clustering algorithm", *International Journal of Production Research*, vol. 18, no. 2, pp. 213–232, 1980.

[KUS 87] KUSIAK A., "The generalized group technology concept", *International Journal of Production Research*, vol. 25, no. 4, pp. 561–569, 1987.

[MAD 14] MADANI G., Analyse de la chaîne de valeur interne et mise en place d'un projet Lean, Masters' thesis, University of Tunis, Tunisia, 2014.

[PIL 11] PILLET M., MARTIN-BONNEFOUS CH., BONNEFOUS P. *et al.*, *Gestion de production: Les fondamentaux et les bonnes pratiques*, Editions d'Organisation, Eyrolles, Paris, 2011.

5.5. Sheet 30: Production Planning: the Business Plan

[JOH 05] JOHNSON G., SCHOLES K., WHITTINGTON R. *et al.*, *Stratégique*, Pearson Education France, Paris, 2005.

[LEA 65] LEARNED E., CHRISTENSEN C.R., ANDREWS K. *et al.*, *Business policy*, Homewood III, Irwin, 1965.

[LEA 69] LEARNED E., CHRISTENSEN C., ANDREWS K. *et al.*, *Business policy, text and cases*, Homewood III, Irwin, 1969.

[POR 79] PORTER M., "How competitive forces shape strategy", *Harvard Business Review*, no. 57, pp. 86–93, March–April 1879.

[POR 86] PORTER M., *L'avantage concurrentiel : comment devancer ses concurrents et maintenir son avance*, Economica, Paris, 1986.

[STU 12] STUTELY R., *Concevoir un Business Plan Efficace*, Pearson Eduction France, Paris, 2012.

6.1. Sheet 33: The Problem of Downstream or Distribution Logics

[BON 99] BONET D., Conflit et coopération dans le canal de distribution : l'analyse du discours des acteurs comme révélateur des comportements stratégiques, PhD thesis, Université de la Méditerranée d'Aix-Marseille II, June 1999.

[BON 05] BONET D., PACHÉ G., "A new approach for understanding hindrances to collaborative practices in the logistics channel", *International Journal of Retail & Distribution Management*, vol. 33, no. 8, pp. 583–596, 2005.

[DES 00] DESGARETS V., "Les relations distributeur-fabricant et les modalités de leur coordination", in FABBE-COSTES N., COLIN J., PACHÉ G. (eds), *Faire de la recherche en logistique et distribution?*, Vuibert-Fnege, Paris, 2000.

[DUR 10] DURAND B., GONZALES-FELIU J., HENRIOT F., "La logistique urbaine, facteur-clé de développement du B to C", *8e Rencontres Internationales de la recherche en Logistique*, Bordeaux, France, 29–30 September and 1 October 2010.

[FEN 05] FENNETEAU H., NARO G., "Contrôle de confiance dans l'entreprise virtuelle: illustrations logistiques", *Revue Française de Gestion*, no. 156, pp. 203–219, 2005.

[LIV 06] LIVOLSI L., CAMMAN C., ROUSSAT C., *Lexipro. Le lexique des termes de la logistique*, Groupe Liaisons, Paris, 2006.

[PAC 97] PACHÉ G., DES GARETS V., "Relations inter-organisationnelles dans les canaux de distribution : les dimensions logistiques", *Recherche et Applications en Marketing*, vol. 12, no. 2, pp. 61–82, 1997.

[PAC 08] PACHÉ G., "Efficient urban e-logistics: mutualization of resources and source of competitive advantage", *7th International Meeting for Research in Logistics*, Avignon, France, 24–26 September 2008.

[VAN 98] VANDAELE M., "L'ECR: moyen indispensable du partenariat producteurs-distributeurs", *Décisions Marketing*, no. 15, pp. 69–76, 1998.

6.2. Sheet 34: The Warehouse/Platform Alternative in Distribution Circuit

[CER 14] CEREMA, Réglementation locale du transport de marchandises en ville, http://www.territoires-ville.cerema.fr/reglementation-locale-du-transport-de-marchandises-a696.html, 2014.

6.3. Sheet 35: Urban Logistics or Last Mile Logistics

[BAL 14] BALLOT E., MONTREUIL B., *L'Internet physique, le réseau des réseaux des prestations logistiques*, La Documentation française, Paris, 2014.

[BAR 13] BARDIN I., MOESCH E., "Le portage d'achats en centre-ville : état de l'existant", *3ème Colloque de Logistique Urbaine*, Nantes, France, Thursday 27 June 2013.

[BEA 99] BEAULIEU M., MARTIN R., LANDRY S., "Logistique à rebours: un portrait nord-américain", *Logistique & Management*, vol. 7, no. 2, pp. 5–14, 1999.

[BOU 06] BOUDOUIN D., *Guide méthodologique : les espaces logistiques urbains*, La Documentation française, Paris, 2006.

[CAR 98] CARTER C., ELLRAM L., "Reverse logistics: a review of the literature and framework for future investigation", *Journal of Business Logistics*, vol. 19, no. 1, pp. 85–102, 1998.

[ELK 10] EL KORCHI A., MILLET D., "Conditions stratégiques d'émergence d'une reverse supply chain fondée sur le remanufacturing chez le fabricant d'origine", *8es Rencontres Internationales de Recherche en Logistique et SCM*, Bordeaux, France, 29–30 September, 1 October 2010.

[FER 17] FERGATI R., L'impression 3D, prochaine révolution dans la logistique industrielle: la stratégie d'UPS, VAE thesis, Université Lumière Lyon 2, September 2017.

[GCD 12] GCDD – GROUPE CAISSE DES DÉPÔTS, Réflexions prospectives Ville 2030, p. 17, http://www.comite21.org/docs/infos21/2014/avril-prospective/reflexions-prospectives-cdc.pdf, 2012.

[GON 13] GONZALEZ-FELIU J., MALHÉNÉ N., MORGANTI E. *et al.*, Développement des espaces logistiques urbains, CDU et ELP dans l'Europe du Sud-Est, https://halshs.archives-ouvertes.fr/halshs-00862009/document, 2013.

[LAM 03] LAMBERT S., RIOPEL D., "Logistique inverse: revue de littérature", *Les Cahiers du GERAD*, no. G-2003-61, October 2003.

[LAM 04] LAMBERT S., RIOPEL D., "Modèle intégrateur de la logistique inverse", *Les Cahiers du GERAD*, no. G-2004-69, August 2004.

[MIC 10] MICHAUD C., Comportements des consommateurs et biens privés à caractéristiques environnementales: une approche expérimentale, PhD thesis, University of Grenoble, March 2010.

[MOR 15] MORANA J., "La logistique urbaine durable de demain", *Logistique & Management*, vol. 23, no. 2, pp. 7–20, 2015.

[PAT 02] PATIER D., *La logistique dans la ville*, Celse, Paris, 2002.

[ROG 99] ROGERS D., TIBBEN-LEMBKE R., "Reverse logistics: stratégies et techniques", *Logistique & Management*, vol. 7, no. 2, pp. 15–25, 1999.

[ROG 01] ROGERS D., TIBBEN-LEMBKE R., "An examination of reverse logistics practices", *Journal of Business Logistics*, vol. 22, no. 2, pp. 129–147, 2001.

7.1. Sheet 36: Traceability

[FAB 06] FABBE-COSTES N., "Traçabilité et logistique : les interactions", *Techniques de l'ingénieur*, 2006.

[KAR 10] KARÂA M., Les déterminants de l'adoption de la traçabilité par les entreprises de conditionnement de dattes en Tunisie, PhD thesis, Université de la Méditerranée Aix-Marseille II, 2010.

[LAZ 14] LAZZERI J., Vers une traçabilité totale des supply chains : le cas de l'agroalimentaire en France, PhD thesis, Aix-Marseille University, 2014.

[MEV 09] MEVEL O., LERAY Y., "Les promesses relatives à l'implémentation d'une solution RFID : le cas d'une centrale d'achat française", *Revue Française de Gestion Industrielle*, vol. 28, no. 2, pp. 53–70, 2009.

[PEL 07] PELLATON A., VIRUEGA J.-L., "L'utilisation de la traçabilité pour la sécurité sanitaire : analyse par la théorie de la traduction", *1ère journée de recherche Relations entre Industrie et Grande Distribution Alimentaire*, Avignon, 29 March 2007.

7.3. Sheet 38: Information and Communication Technologies and Information Systems

[AGO 11] AGORA FONCTIONS, La Supply Chain : une fonction stratégique en action, 2011.

[EVR 11] EVRARD-SAMUEL K., RUEL S., SPALANZANI A., "Systèmes d'information et résilience des chaînes logistiques globales : Proposition d'un écosystème informationnel", *Cahier de recherche du CERAG no. 2011-02 E5*, 2011.

[LAU 10] LAUDON K., LAUDON J., ELRAGAL A., *Management Information Systems*, Pearson Education Limited, London, 2010.

[ROG 15] ROGNON J.-L., "Comment tirer le meilleur parti de son projet ERP?", *Supply Chain Magazine*, no. 99, pp. 88–98, available at: http://supplychainmagazine.fr/TOUTE-INFO/Archives/SCM099/DOSSIER-99.pdf, November 2015.

7.4. Sheet 39: Quality Implements and Logistics

[CHA 00] CHAQUAL X., AMDEC, available at: http://chaqual.free.fr/outils/amdec/amdec.html, 2000.

[FOR 97] FORKER L., MENDEZ D., HERSHAUER J., "Total quality management in the supply chain: What is its impact on performance?", *International Journal of Production Research*, vol. 35, no. 6, pp. 1681–1701, 1997.

[JUR 99] JURAN J., GODFREY A., *Juran's Quality Handbook*, McGraw-Hill, New York, 1999.

[LAM 99] LAMOURI S., THOMAS A., "Juste à temps et qualité totale : concepts et outils", *Techniques de L'ingénieur*, AG 5 190, 1999.

[LED 01] LE DENN Y., *La chaîne logistique au service du client*, Celse, Paris, 2001.

[MIL 80] MIL-STD-1629A (1980), Military Standard – Procedures for performing a failure mode, effects and criticality analysis, available at: https://src.alionscience.com/pdf/MIL-STD-1629RevA.pdf, 24 November 1980.

7.6. Sheet 41: Measuring Performance in Logistics

[CHI 94] CHIAPELLO E., DELMOND M.H., "Les tableaux de bord de gestion, outils d'introduction du changement", *Revue Française de Gestion*, no. 97, pp. 49–58, 1994.

[KAP 92] KAPLAN R., NORTON D., "The balanced scorecard – measures that drives performance", *Harvard Business Review*, vol. 70, no. 1, pp. 71–79, 1992.

[KAP 96] KAPLAN R., NORTON D., *The Balanced Scorecard*, Harvard Business School Press, Boston, 1996.

[MOR 13] MORANA J., *Sustainable Supply Chain Management*, ISTE, London and John Wiley & Sons, New York, 2013.

[MOR 15] MORANA J., GONZALEZ-FELIU J., "A sustainable urban logistics dashboard from the perspective of a group of logistics managers", *Management Research Review,* vol. 38, no. 10, pp. 1068–1085, 2015.

C.1. Sheet 42: Conclusion – Some Atypical Logistics

[AND 07] ANDRÉ V., FÉNIES P., "Modélisation et simulation des flux logistiques du Nouvel Hôpital d'Estaing", *Logistique & Management*, vol. 15, no. 1, pp. 49–59, 2007.

[CHA 06] CHANDES J., PACHÉ G., "La coordination des chaînes logistiques multi-acteurs dans un contexte humanitaire : quels cadres conceptuels pour améliorer l'action ?", *Logistique & Management*, vol. 14, no. 1, pp. 33–42, 2006.

[CHA 07] CHAABANE S., MESKENS N., GUINET A., "Comparaison des performances des politiques de programmation opératoire", *Logistique & Management*, vol. 15, no. 1, pp. 17–26, 2007.

[FÉN 04] FÉNIES P., GOURGAND M., TCHERNEV N., "Une contribution à la mesure de la performance de la Supply Chain hospitalière : l'exemple du processus opératoire", *2ème conférence francophone en Gestion et Ingénierie de Systèmes Hospitaliers (GISEH)*, Mons, Belgium, 9–11 September 2004.

[FÉN 05] FÉNIES P., TCHERNEV N., "La modélisation d'une unité générique de soins : une brique essentielle pour le système d'information et d'aide à la décision de la supply chain du nouvel hôpital d'Estaing", *Logistique & Management*, vol. 13, no. 2, pp. 45–58, 2005.

[GIR 09] GIROUX H., BEAULIEU M., COOREN F., "Gérer les chaînes logistiques humanitaires : l'expérience de Médecins Sans Frontières", *Logistique & Management,* vol. 17, no. 1, pp. 65–76, 2009.

[LAN 00] LANDRY S., BEAULIEU M., FRIEL T. *et al.*, Etude internationale des meilleures pratiques de logistique hospitalière, Groupe de Recherche CHAINE, cahier nos 00–05, 2000.

[LAN 02] LANDRY S., BEAULIEU M., "La logistique hospitalière : un remède aux maux du secteur de la santé ?", *Gestion: Revue Internationale de Gestion*, vol. 26, no. 4, pp. 34–41, 2002.

[LIÈ 09] LIÈVRE P., GAUTIER A., "Les registres de la logistique des situations extrêmes : des expéditions polaires aux services d'incendies et secours", *Revue Management & Avenir*, vol. 4, no. 24, pp. 196–216, 2009.

[MON 11] MONCHANIN J., "Le CHU de Dijon adopte le Kanban", *Stratégies Logistique*, no. 129, pp. 24–26, June–July 2011.

[MOU 13] MOULY B., "Le CHU de Nantes dope sa préparation de commandes", *Logistiques Magazine*, no. 281, pp. 56–58, July–August 2013.

[RIV 98] RIVOLIER J., "Introduction", *Bulletin de psychologie*, vol. 51, no. 6, pp. 717–718, 1998.

[SAL 16] SALAUN V., "Demain, quel management logistique pour la logistique événementielle ? Une étude prospective sur l'impact du cashless sur le management logistique des festivals musicaux en France à l'horizon 2020", *11e Rencontres Internationales de la Recherche en Logistique (RIRL 2016)*, Lausanne, Switzerland, September 2016.

[THO 03] THOMAS A., "Why logistics?", *Forced Migration Review*, no. 18, p. 4, 2003.

[TRU 05] TRUNICK P., "Special report: delivering relief to Tsunami victims", *Logistics Today*, vol. 46, no. 2, pp. 1–3, 2005.

[UNI 17] UNIMEV – UNION FRANÇAISE DES MÉTIERS DE L'EVÈNEMENT, Données générales et retombées économiques des Foires, Salons et Congrès en France et Ile-de-France, p. 4, available at: http://www.unimev.fr/ files/unimev.fr/public/ ressources/files/memo_chiffres_2017.pdf, 31 March 2017.

[VAU 09] VAUCLARE C., "Les événements culturels : essai de typologie", *Culture études*, vol. 3, no. 3, pp. 1–8, 2009.

Index

U, V, W

Other titles from

ISTE

in

Systems and Industrial Engineering – Robotics

2018

BERRAH Lamia, CLIVILLÉ Vincent, FOULLOY Laurent
Industrial Objectives and Industrial Performance: Concepts and Fuzzy Handling

GONZALEZ-FELIU Jesus
Sustainable Urban Logistics: Planning and Evaluation

GROUS Ammar
Applied Mechanical Design

LEROY Alain
Production Availability and Reliability: Use in the Oil and Gas Industry

MARÉ Jean-Charles
Aerospace Actuators 3: European Commercial Aircraft and Tiltrotor Aircraft

MAXA Jean-Aimé, BEN MAHMOUD Mohamed Slim, LARRIEU Nicolas
Model-driven Development for Embedded Software: Application to Communications for Drone Swarm

MBIHI Jean
Analog Automation and Digital Feedback Control Techniques
Advanced Techniques and Technology of Computer-Aided Feedback Control

SIMON Christophe, WEBER Philippe, SALLAK Mohamed
Data Uncertainty and Important Measures
(Systems Dependability Assessment Set – Volume 3)

TANIGUCHI Eiichi, THOMPSON Russell G.
City Logistics 1: New Opportunities and Challenges

TANIGUCHI Eiichi, THOMPSON Russell G.
City Logistics 2: Modeling and Planning Initiatives

TANIGUCHI Eiichi, THOMPSON Russell G.
City Logistics 3: Towards Sustainable and Liveable Cities

2017

ANDRÉ Jean-Claude
From Additive Manufacturing to 3D/4D Printing 1: From Concepts to Achievements
From Additive Manufacturing to 3D/4D Printing 2: Current Techniques, Improvements and their Limitations
From Additive Manufacturing to 3D/4D Printing 3: Breakthrough Innovations: Programmable Material, 4D Printing and Bio-printing

ARCHIMÈDE Bernard, VALLESPIR Bruno
Enterprise Interoperability: INTEROP-PGSO Vision

CAMMAN Christelle, FIORE Claude, LIVOLSI Laurent, QUERRO Pascal
Supply Chain Management and Business Performance: The VASC Model

FEYEL Philippe
Robust Control, Optimization with Metaheuristics

MARÉ Jean-Charles
Aerospace Actuators 2: Signal-by-Wire and Power-by-Wire

POPESCU Dumitru, AMIRA Gharbi, STEFANOIU Dan, BORNE Pierre
Process Control Design for Industrial Applications

RÉVEILLAC Jean-Michel
Modeling and Simulation of Logistics Flows 1: Theory and Fundamentals
Modeling and Simulation of Logistics Flows 2: Dashboards, Traffic Planning and Management
Modeling and Simulation of Logistics Flows 3: Discrete and Continuous Flows in 2D/3D

2016

ANDRÉ Michel, SAMARAS Zissis
Energy and Environment
(Research for Innovative Transports Set - Volume 1)

AUBRY Jean-François, BRINZEI Nicolae, MAZOUNI Mohammed-Habib
Systems Dependability Assessment: Benefits of Petri Net Models (Systems Dependability Assessment Set - Volume 1)

BLANQUART Corinne, CLAUSEN Uwe, JACOB Bernard
Towards Innovative Freight and Logistics (Research for Innovative Transports Set - Volume 2)

COHEN Simon, YANNIS George
Traffic Management (Research for Innovative Transports Set - Volume 3)

MARÉ Jean-Charles
Aerospace Actuators 1: Needs, Reliability and Hydraulic Power Solutions

REZG Nidhal, HAJEJ Zied, BOSCHIAN-CAMPANER Valerio
Production and Maintenance Optimization Problems: Logistic Constraints and Leasing Warranty Services

TORRENTI Jean-Michel, LA TORRE Francesca
Materials and Infrastructures 1 (Research for Innovative Transports Set - Volume 5A)
Materials and Infrastructures 2 (Research for Innovative Transports Set - Volume 5B)

WEBER Philippe, SIMON Christophe
Benefits of Bayesian Network Models
(Systems Dependability Assessment Set – Volume 2)

YANNIS George, COHEN Simon
Traffic Safety (Research for Innovative Transports Set - Volume 4)

2015

AUBRY Jean-François, BRINZEI Nicolae
Systems Dependability Assessment: Modeling with Graphs and Finite State Automata

BOULANGER Jean-Louis
CENELEC 50128 and IEC 62279 Standards

BRIFFAUT Jean-Pierre
E-Enabled Operations Management

MISSIKOFF Michele, CANDUCCI Massimo, MAIDEN Neil
Enterprise Innovation

2014

CHETTO Maryline
Real-time Systems Scheduling
Volume 1 – Fundamentals
Volume 2 – Focuses

DAVIM J. Paulo
Machinability of Advanced Materials

ESTAMPE Dominique
Supply Chain Performance and Evaluation Models

FAVRE Bernard
Introduction to Sustainable Transports

GAUTHIER Michaël, ANDREFF Nicolas, DOMBRE Etienne
Intracorporeal Robotics: From Milliscale to Nanoscale

MICOUIN Patrice
Model Based Systems Engineering: Fundamentals and Methods

MILLOT Patrick
Designing Human–Machine Cooperation Systems

NI Zhenjiang, PACORET Céline, BENOSMAN Ryad, RÉGNIER Stéphane
Haptic Feedback Teleoperation of Optical Tweezers

OUSTALOUP Alain
Diversity and Non-integer Differentiation for System Dynamics

REZG Nidhal, DELLAGI Sofien, KHATAD Abdelhakim
Joint Optimization of Maintenance and Production Policies

STEFANOIU Dan, BORNE Pierre, POPESCU Dumitru, FILIP Florin Gh., EL KAMEL Abdelkader
Optimization in Engineering Sciences: Metaheuristics, Stochastic Methods and Decision Support

2013

ALAZARD Daniel
Reverse Engineering in Control Design

ARIOUI Hichem, NEHAOUA Lamri
Driving Simulation

CHADLI Mohammed, COPPIER Hervé
Command-control for Real-time Systems

DAAFOUZ Jamal, TARBOURIECH Sophie, SIGALOTTI Mario
Hybrid Systems with Constraints

FEYEL Philippe
Loop-shaping Robust Control

FLAUS Jean-Marie
Risk Analysis: Socio-technical and Industrial Systems

FRIBOURG Laurent, SOULAT Romain
Control of Switching Systems by Invariance Analysis: Application to Power Electronics

GROSSARD Mathieu, REGNIER Stéphane, CHAILLET Nicolas
Flexible Robotics: Applications to Multiscale Manipulations

GRUNN Emmanuel, PHAM Anh Tuan
Modeling of Complex Systems: Application to Aeronautical Dynamics

HABIB Maki K., DAVIM J. Paulo
Interdisciplinary Mechatronics: Engineering Science and Research Development

HAMMADI Slim, KSOURI Mekki
Multimodal Transport Systems

JARBOUI Bassem, SIARRY Patrick, TEGHEM Jacques
Metaheuristics for Production Scheduling

KIRILLOV Oleg N., PELINOVSKY Dmitry E.
Nonlinear Physical Systems

LE Vu Tuan Hieu, STOICA Cristina, ALAMO Teodoro, CAMACHO Eduardo F., DUMUR Didier
Zonotopes: From Guaranteed State-estimation to Control

MACHADO Carolina, DAVIM J. Paulo
Management and Engineering Innovation

MORANA Joëlle
Sustainable Supply Chain Management

SANDOU Guillaume
Metaheuristic Optimization for the Design of Automatic Control Laws

STOICAN Florin, OLARU Sorin
Set-theoretic Fault Detection in Multisensor Systems

2012

AÏT-KADI Daoud, CHOUINARD Marc, MARCOTTE Suzanne, RIOPEL Diane
Sustainable Reverse Logistics Network: Engineering and Management

BORNE Pierre, POPESCU Dumitru, FILIP Florin G., STEFANOIU Dan
Optimization in Engineering Sciences: Exact Methods

CHADLI Mohammed, BORNE Pierre
Multiple Models Approach in Automation: Takagi-Sugeno Fuzzy Systems

DAVIM J. Paulo
Lasers in Manufacturing

DECLERCK Philippe
Discrete Event Systems in Dioid Algebra and Conventional Algebra

DOUMIATI Moustapha, CHARARA Ali, VICTORINO Alessandro, LECHNER Daniel
Vehicle Dynamics Estimation using Kalman Filtering: Experimental Validation

GUERRERO José A, LOZANO Rogelio
Flight Formation Control

HAMMADI Slim, KSOURI Mekki
Advanced Mobility and Transport Engineering

MAILLARD Pierre
Competitive Quality Strategies

MATTA Nada, VANDENBOOMGAERDE Yves, ARLAT Jean
Supervision and Safety of Complex Systems

POLER Raul *et al.*
Intelligent Non-hierarchical Manufacturing Networks

TROCCAZ Jocelyne
Medical Robotics

YALAOUI Alice, CHEHADE Hicham, YALAOUI Farouk, AMODEO Lionel
Optimization of Logistics

ZELM Martin *et al.*
Enterprise Interoperability –I-EASA12 Proceedings

2011

CANTOT Pascal, LUZEAUX Dominique
Simulation and Modeling of Systems of Systems

DAVIM J. Paulo
Mechatronics

DAVIM J. Paulo
Wood Machining

GROUS Ammar
Applied Metrology for Manufacturing Engineering

KOLSKI Christophe
Human–Computer Interactions in Transport

LUZEAUX Dominique, RUAULT Jean-René, WIPPLER Jean-Luc
Complex Systems and Systems of Systems Engineering

ZELM Martin, *et al.*
Enterprise Interoperability: IWEI2011 Proceedings

2010

BOTTA-GENOULAZ Valérie, CAMPAGNE Jean-Pierre, LLERENA Daniel, PELLEGRIN Claude
Supply Chain Performance / Collaboration, Alignement and Coordination

BOURLÈS Henri, GODFREY K.C. Kwan
Linear Systems

BOURRIÈRES Jean-Paul
Proceedings of CEISIE'09

CHAILLET Nicolas, REGNIER Stéphane
Microrobotics for Micromanipulation

DAVIM J. Paulo
Sustainable Manufacturing

GIORDANO Max, MATHIEU Luc, VILLENEUVE François
Product Life-Cycle Management / Geometric Variations

LOZANO Rogelio
Unmanned Aerial Vehicles / Embedded Control

LUZEAUX Dominique, RUAULT Jean-René
Systems of Systems

VILLENEUVE François, MATHIEU Luc
Geometric Tolerancing of Products

2009

DIAZ Michel
Petri Nets / Fundamental Models, Verification and Applications

OZEL Tugrul, DAVIM J. Paulo
Intelligent Machining

PITRAT Jacques
Artificial Beings

2008

ARTIGUES Christian, DEMASSEY Sophie, NERON Emmanuel
Resources–Constrained Project Scheduling

BILLAUT Jean-Charles, MOUKRIM Aziz, SANLAVILLE Eric
Flexibility and Robustness in Scheduling

DOCHAIN Denis
Bioprocess Control

LOPEZ Pierre, ROUBELLAT François
Production Scheduling

THIERRY Caroline, THOMAS André, BEL Gérard
Supply Chain Simulation and Management

2007

DE LARMINAT Philippe
Analysis and Control of Linear Systems

DOMBRE Etienne, KHALIL Wisama
Robot Manipulators

LAMNABHI Françoise *et al.*
Taming Heterogeneity and Complexity of Embedded Control

LIMNIOS Nikolaos
Fault Trees

2006

FRENCH COLLEGE OF METROLOGY
Metrology in Industry

NAJIM Kaddour
Control of Continuous Linear Systems

Printed and bound by CPI Group (UK) Ltd, Croydon, CR0 4YY

15/09/2026

14962349-0002